Adhesion

Its Role in Inflammatory Disease

Breakthroughs in Molecular Biology

Adhesion: Its Role in Inflammatory Disease is the fourth volume to appear in this exciting new series of high quality, affordable books in the fields of molecular biology and immunology. This series is dedicated to the rapid publication of the latest breakthroughs and cutting edge technologies as well as syntheses of major advances within molecular biology.

Other volumes in the series include:

PCR Technology: Principles and Applications for DNA Amplification
edited by H. Erlich

DNA Fingerprinting: An Introduction
by L. T. Kirby

Antibody Engineering: A Practical Guide
edited by C.A.K. Borrebaeck

Adhesion

Its Role in Inflammatory Disease

John M. Harlan
and
David Y. Liu
Editors

W. H. Freeman and Company

New York

Library of Congress Cataloging-in-Publication Data

Adhesion : its role in inflammatory disease / edited by John M. Harlan & David Y. Liu.
    p. cm.
    Includes index.
    ISBN 0-7167-7010-5
    1. Inflammation--Pathophysiology. 2. Cell adhesion. 3. Cell adhesion molecules.
    I. Harlan, John M. II. Liu, David Y. 1950-
    [DNLM: 1. Cell Adhesion--physiology. 2. Cell Adhesion Molecules--physiology. 3. Endothelium--physiology. 4. Inflammation--physiopathology. 5. Leukocytes--physiology. QZ 150 H234]
RB131.A34 1992
616'.0473--dc20
DNLM/DLC
for Library of Congress
                                  91-42770
                                  CIP

Copyright © 1992 by W.H. Freeman and Company

No part of this book may be reproduced by any mechanical, photographic, or electronic process, or in the form of a phonographic recording, nor may it be stored in a retrieval system, transmitted, or otherwise copied for public or private use, without written permission from the publisher.

Printed in the United States of America

1 2 3 4 5 6 7 8 9 0   V B   9 9 8 7 6 5 4 3 2 1

Contributors

Claire M. Doerschuk
Indiana University Medical Center
Indianapolis, Indiana

Carl G. Figdor
The Netherlands Cancer Institute
Amsterdam, The Netherlands

Jennifer R. Gamble
Hanson Centre for Cancer Research
Institute of Medical and Veterinary Science
Adelaide, South Australia

John M. Harlan
University of Washington
Seattle, Washington

Zehra Kaymakcalan
Cetus Corp.
Emeryville, California

Laurence A. Lasky
Genentech, Inc.
South San Francisco, California

David Y. Liu
Cetus Corp.
Emeryville, California

Roy R. Lobb
Biogen, Inc.
Cambridge, Massachusetts

James C. Paulson
Cytel Corp. and Scripps Research Institute
San Diego, California

Charles L. Rice
University of Washington
Seattle, Washington

C. Wayne Smith
Baylor College of Medicine
Houston, Texas

William B. Smith
Hanson Centre for Cancer Research
Institute of Medical and Veterinary Science
Adelaide, South Australia

Mathew A. Vadas
Hanson Centre for Cancer Research
Institute of Medical and Veterinary Science
Adelaide, South Australia

Yvette van Kooyk
The Netherlands Cancer Institute
Amsterdam, The Netherlands

Nicholas B. Vedder
University of Washington
Seattle, Washington

Robert K. Winn
University of Washington
Seattle, Washington

Contents

Preface . . . . **xi**

1. Integrin–Immunoglobulin Superfamily Interactions in Endothelial–Leukocyte Adhesion **1**

    ICAM/LFA-1 Interactions 2
    VCAM-1/VLA-4 Interaction 7
    Summary 13
    References 15

2. Selectin/Carbohydrate-Mediated Adhesion of Leukocytes . . **19**

    The Selectin Family—A Homologous Gene Family 20
        The Lectin Domains 20
        CR and EGF Domains 21
        Soluable Forms of CD62 21
    Cellular Distribution and Expression of the Selectins 22
    Carbohydrate Ligands of the Selectins 24
        Mel-14/LAM-1 24
        ELAM-1 25
        SLe^X as a Ligand for ELAM-1 25
        Vim-2 Antigen as a Ligand 26
        CD62 28
    Distribution and Biosynthesis of SLe^X 28
        Identification of Le^X and SLe^X as Leukocyte Antigens 28
        Expression of SLe^X on Glycoproteins and Glycolipids of Leukocytes . . 29
        Biosynthesis of SLe^X 30
        Relevance of SLe^X Expression to the Cell Selectivities of ELAM-1 and CD62 32
    Roles of Selectins in Leukocyte Recruitment *In Vivo* 33
        Mediation of Early Interactions of Leukocytes With Endothelial Cells . 33

ELAM-1 Expression Associated With Acute and Chronic
Inflammatory Disease 36
Summary 37
References 38

3. The Homing Receptor (LECAM 1/L Selectin): A Carbohydrate-Binding Mediator of Adhesion in the Immune System . . . 43

Historical Perspective 44
cDNA Cloning of the Murine and Human Homing Receptor 46
Genomic Structure and Chromosomal Localization of the Human
and Murine Homing Receptor Genes 50
The Nature of the Ligands for the Homing Receptor 51
Structure and Function of Homing Receptor Domains 54
The Neutrophil Homing Receptor: A Mediator of Neutrophil
Rolling *In Vivo* 55
Regulatory Aspects of Homing Receptor Function 58
Summary 60
References 61

4. Regulation of Myeloid Blood Cell–Endothelial Interaction by Cytokines 65

Blood Cell Endothelial Interactions and Their Measurement 65
Cytokines and Adhesion 66
Activation of Endothelium by Cytokines 66
Effect of Cytokines on Endothelial Adhesion Structures 67
Inhibition of Endothelial Activation 68
Transforming Growth Factor-β (TGF-β) 69
IL-4 70
Activation of Blood Cells by Cytokines 70
Neutrophils 71
Eosinophils 76
Monocytes 76
Summary 77
References 78

5. Transendothelial Migration 83

Transendothelial Migration *In Vitro* 83
Transmigration Induced by Cytokine Stimulation of the Endothelium . . 87
Cytokine-Stimulated Endothelial Cells Activate Neutrophil Motility . . . 88
Contributors of Adherence to Transendothelial Migration 89
Specific Adhesion Molecules and Migration Toward Chemotactic Gradients . . 91
Specific Adhesion Molecules and Migration Induced by Cytokine
Stimulation of the Endothelial Cells 94
Contribution of Margination to Neutrophil Localization 98

Three Stages in the Process of Leukocyte Extravasation 103
Specific Adhesion Molecules and the Migration of Lymphocytes
and Other Leukocytes 104
References 107

6. *In Vivo* Models of Leukocyte Adherence to Endothelium . . 117

Intravital Microscopy 118
   Leukocyte Sticking 118
   Leukocyte Rolling 122
   The Marginated Pool and Margination 123
Emigration 125
   CD11/CD18 126
   ICAM-1 129
   L-Selectin 129
   E-Selectin 131
   P-Selectin 132
   VLA-4 and VCAM-1 132
Genetic Deficiency of Adhesion Proteins 133
   Anti-Adhesion Therapy 133
   Efficacy 134
   Safety 141
   Future Directions 144
Conclusion 144
References 144

7. Regulation of Cell Adhesion 151

Mechanisms That Regulate Cell Adhesion 152
Methods of Regulation of Adhesion Receptor Expression 153
   Cytokines 154
   Infectious Agents Can Modulate Adhesion Receptor Expression . . . 158
   Genetic Factors 158
Activation of Adhesion Receptors ("Inside-Out" Signalling) 159
   Antigen-Specific and Nonspecific Activation of LFA-1 159
   Activation of LFA-1 Compared to Other Integrins 162
   Intracellular Messengers and Affinity Modulation of
   Integrin Receptors 162
   Activation Epitopes, the Role of Cations and Receptor
   Conformation 164
Signalling Through Adhesion Receptors ("Outside-In" Signalling) . . . 170
Organization of Adhesion Receptors at the Cell Surface and the
Role of the Cytoskeleton 172
Redundancy 173
Adhesion Cascades 174
Synopsis and Discussion 175
References 176

8. Outlook for the Future 183
   References 186

Appendix - *In Vitro* Adhesion Assay 189
   Endothelial Cells 190
   Leukocyte/Endothelial Cell Static Adhesion Assay 190
   Leukocyte/Endothelial Cell Shear Stress Adhesion Assay 191
   Leukocyte/Recombinant Protein Adhesion Assay 191
   Transendothelial Migration 192
   Detection Methods for Adherent Leukocytes 192
   References 192

Glossary 195

Index 197

Preface

Intercellular adhesion is an important early step in the complex but coordinated sequence of events leading to acute and chronic inflammatory diseases. Understanding the mechanisms of leukocyte binding to endothelial cells will hopefully provide a basis for the eventual successful treatment of inflammation. *Adhesion: Its Role in Inflammatory Disease* is a theoretical and practical compendium of all aspects of leukocyte adhesion (with a particular emphasis on neutrophils) to endothelial cells. The intention is to introduce a complex issue to a broader audience than was previously addressed in more specialized reviews and to discuss in a comprehensive format the most complete and current understanding of the field. This area of scientific investigation is progressing so quickly that a rapidly published book will be very informative. Included are discussions of the structural and functional nature of cellular adhesion molecules, which can be categorized into three supergene families: integrin, immunoglobulin, and selectin. A chapter is devoted to the recent studies elucidating their function in signal transduction. This book provides a unique description of adhesion as it relates to the subsequent event of transendothelial migration of leukocytes. There is a section on soluble mediators that influence the ability of cells to be activated. Furthermore, this book extends its detailed description of adhesion as an *in vitro* phenomenon to the latest studies of *in vivo* animal models, which provide the first significant evidence of adhesion as an essential step in the progression of inflammatory disease. This collection of important advances in the field of intercellular adhesion will certainly be of value not only to graduate students and research investigators but also to the clinical investigator interested in a complete discussion of leukocyte–endothelial cell adhesion and its role in inflammatory disease.

David Y. Liu
Emeryville, California

John M. Harlan
Seattle, Washington

October 1991

CHAPTER 1

Integrin–Immunoglobulin Superfamily Interactions in Endothelial–Leukocyte Adhesion

Roy R. Lobb

The interaction of circulating leukocytes with the endothelium during immune and inflammatory reactions is dependent upon a series of transient cellular adhesive events. In recent years, significant progress has been made in the characterization of these adhesive interactions at the molecular level.[1-8] Three general mechanisms of adhesion have now been defined: rapid transient upregulation of leukocyte adhesion molecules (e.g., CD11b/CD18);[1] rapid transient upregulation of endothelial adhesion molecules (e.g., granule membrane protein-140 [GMP-140]; and more persistent cytokine activation of endothelial cells, leading to the synthesis of new adhesion molecules (e.g. endothelial leukocyte adhesion molecule-1 [ELAM-1]). In addition, three superfamilies of adhesion molecules have emerged: the selectins or lectin-epidermal growth factor-complement related cell adhesion molecules (LECCAMs) (e.g., ELAM-1 and GMP-140), which appear to interact with carbohydrate-containing ligands; the heterodimeric integrins (e.g., lymphocyte function-associated antigen-1 [LFA-1] and very late antigen-4 [VLA-4]); and members of the immunoglobulin (Ig) superfamily. The characterization of one ligand/receptor pair involved in endothelial–leukocyte adhesion, intercellular adhesion molecule-1 (ICAM-1) and LFA-1, established an unexpected intersection of the immunoglobulin (Ig) and integrin superfamilies.[9] Subsequent studies identified a second Ig/integrin superfamily interaction, that of vascular cell adhesion molecule-1 (VCAM-1) and VLA-4.[10,11] These two adhesion pathways will be discussed in relation to leukocyte adhesion to vascular endothelium *in vitro* and *in vivo*.

ICAM/LFA-1 INTERACTIONS

The leukocyte integrins are a subgroup of the integrin superfamily and play a major role in leukocyte adhesion. They share a common beta subunit (CD18) that can associate noncovalently with one of three different alpha subunits, designated CD11a/LFA1a, CD11b/Mac1a, and CD11c/p150, 95a (Figure 1-1). Of the three integrins, lymphocyte function-associated antigen (LFA-1) (CD11a/CD18) has been the most extensively studied, and its importance in adhesion is the best understood. Its role was first demonstrated using monoclonal antibodies (MAbs) to LFA-1, which were found to block numerous lymphocyte functions, including their adhesion to human endothelial cells.[12-15]

ICAM-1 was identified as a putative LFA-1 ligand expressed on B-lymphoblastoid cells, and independently as a B cell activation marker.[16-18] The role of ICAM-1 in LFA-1-dependent adhesion was defined using MAb inhibition of homotypic aggregation of B-lymphoblastoid cells.[16] MAbs identified in this way have been used to show the interaction of ICAM-1 with LFA-1 in numerous systems and to characterize the antigen. ICAM-1 is a single chain polypeptide of 55 kDa, expressed as a glycoprotein of 76–114 kDa on different cells due to differential glycosylation. ICAM-1 is weakly expressed on peripheral blood leukocytes but its expression on mononuclear leukocytes is increased by activation.[15] In addition, its expression on nonhematopoietic cells is upregulated by cytokines. This is exemplified by its upregulation on endothelium *in vitro* and *in vivo* by interleukin-1 (IL-1), tumor necrosis factor (TNF), and interferon-γ (IFN-g).[14,19,20]

ICAM-1 was cloned independently by two groups.[21,22] The cloning revealed that ICAM-1 was a member of the immunoglobulin superfamily. Members of this superfamily are characterized by the presence of one or more Ig homology regions, each consisting of a disulfide-bridged loop that has a number of antiparallel β-pleated strands arranged in two sheets.[23,24] Three types of Ig homology regions (C, V, and H, also known as C1, V, and C2) have been defined, each with a typical length and consensus of amino acid residues at key positions relative to the two cysteines that form the disulfide link.[23,24] ICAM-1 has five Ig-like domains, designated D1 through D5, beginning at the N terminus (Figure 1-2). All five domains are of the C2 or H type, which are found in other Ig superfamily members known to have adhesive

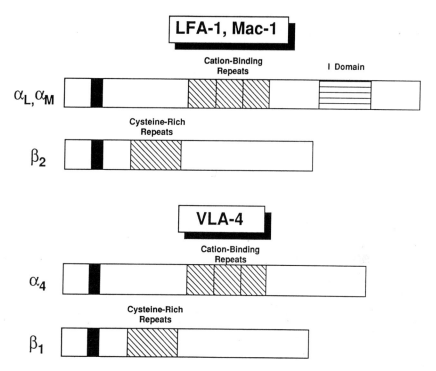

Figure 1-1. Schematic showing domain structures of the integrin superfamily members LFA-1, Mac-1, and VLA-4. LFA-1 and Mac-1 contain an inserted domain, or I domain, absent from VLA-4. All three alpha chains contain three cation-binding repeats, and both beta chains contain a region of cysteine-rich repeats.

functions. ICAM-1 is most closely related to neural cell adhesion molecule (NCAM) and myelin-associated glycoprotein (MAG), two adhesion proteins of the adult nervous system, both of which have five Ig-like domains. ICAM-1 is 24% homologous with MAG and 20% homologous with NCAM over the five Ig-like domains.[21,22]

The cloning of ICAM-1 has allowed detailed structure/function studies to be performed.[25] Deletion of entire domains by oligonucleotide-directed mutagenesis, followed by transient expression of the mutants in COS cells, defined those domains important for LFA-1-dependent adhesion. Deletion of D3, D5, D4 and D5, or the cytoplasmic tail of ICAM-1 had no significant effect on LFA-1-dependent adhesion. Deletion of D3, D4, and D5 combined reduced LFA-1-dependent adhesion about three-fold.[25] In further studies, the Ig-like structure of ICAM-1 was used as a guide to introduce amino acid substitutions into the loops connecting the putative β-strands of the first three domains. Mutations with the strongest effect on LFA-1 adhesion were localized to D1. A single substitution, E34/A, completely eliminated LFA-1

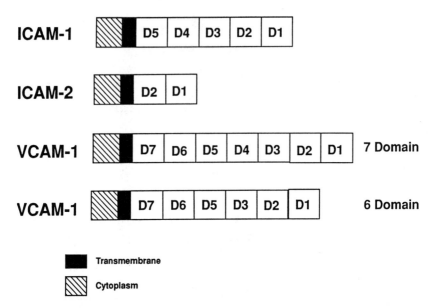

Figure 1-2. Schematic showing domain structures of the Ig superfamily members ICAM-1, ICAM-2, and the two alternatively spliced forms of VCAM-1.

binding, while a second, Q73/H, reduced it by an order of magnitude. Other mutations in both D1 and D2 had two-fold or lower effects, while mutations in D3 were without effect. The results show that the amino-terminal two domains of ICAM-1 alone are sufficient to support functional interactions with LFA-1, while D1 appears to be most critical for LFA-1 binding.[25]

The majority of rhinoviruses, which cause about 50% of common colds, and some Coxsackie viruses, share a common receptor on human cells, and this receptor has recently been identified as ICAM-1.[26-28] Examination of ICAM-1 deletion mutants and amino acid substitution mutants shows that the first two domains support viral adhesion and that D1 is more important than D2, but that the viral and LFA-1 binding sites only partially overlap. Interestingly, the virus will only bind to human, not murine, ICAM-1, and viral adhesion is cation-independent.[25] In addition, a recombinant soluble form of ICAM-1 blocks rhinovirus infection *in vitro*.[29] ICAM-1 is one of several Ig superfamily adhesion molecules that have been subverted by viruses as receptors for cell infectivity. Other examples include the human immunodeficiency virus (HIV) receptor, CD4, and the poliovirus receptor.[30] Whether VCAM-1 is also used as a viral receptor is presently unknown. ICAM-1 is also one of two receptors characterized for *Plasmodium falciparum*-infected erythrocytes.[31] A key event in malaria is the sequestration of infected erythrocytes in the cerebral microvasculature, and ICAM-1 may play a role in this process.

Recent studies show that both ICAM-1 and LFA-1 are ligands for other adhesion molecules. Functional evidence for a second ligand for LFA-1 was obtained from blocking MAb studies. While most LFA-1-dependent functions could also be blocked by MAbs to ICAM-1, several could not, including LFA-1-dependent aggregation of the T-lymphoblastoid line SKW3, and importantly, lymphocyte adhesion to endothelial cells.[14,16] In an elegant series of experiments, expression cloning was used to identify a second ligand for LFA-1.[32] The cDNA was obtained from a human endothelial cell library and was designated ICAM-2. Like ICAM-1, ICAM-2 is a member of the Ig superfamily, but contains only two Ig-like domains that are most homologous to the two N-terminal domains of ICAM-1, showing 35% identity (see Figure 1–2). Based on Northern blot data, ICAM-2 appears to be a noninducible LFA-1-dependent adhesion molecule, present in human umbilical vein endothelial cells (HUVECs) and various mononuclear leukocytic cell lines.[32]

Initially, studies on ICAM-2 were hampered by lack of specific MAbs, particularly those able to block ICAM-2 function. However, two groups have recently generated blocking MAbs to ICAM-2, allowing rapid progress to be made in its further characterization.[33,34] One group generated MAbs using COS cells transiently expressing ICAM-2 as an immunogen.[33] One of the MAbs, designated CBR-IC2/2, was found to block ICAM-2-dependent adhesion to LFA-1 completely. Immunoprecipitations from several cell types revealed a glycosylated protein of about 55–65 kD under reducing conditions. Using this MAb, the tissue distribution of ICAM-2 was examined and ICAM-2 was found restricted largely to endothelium and certain interstitial cells. Flow cytometric analysis of leukocytes shows low uninducible expression of ICAM-2 on mononuclear leukocytes, including natural killer NK cells, and its absence from PMN. ICAM-2 is also present at moderate levels on a variety of T- and B-lymphoblastoid cell lines. It is most intensely expressed *in vitro* on HUVECs. The data obtained with MAbs parallels the information suggested from Northern blot analyses.[32] Importantly, functional studies show that ICAM-1 and ICAM-2 account for all LFA-1-dependent adhesion to either resting or activated endothelium.[33]

A second group has generated blocking MAbs to ICAM-2 using as an immunogen a protein A-ICAM-2 fusion protein expressed in *Escherichia coli*.[34] The MAb, designated 6D5, immunoprecipitates a 55 kD glycoprotein from endothelial cells and transiently transfected COS cells, and flow cytometric analyses show a similar distribution of ICAM-2 on lymphoblastoid cell lines to that described above. MAb 6D5 also blocks adhesion completely and binds to the most N-terminal domain of ICAM-2.

The constitutive noninducible high expression levels of ICAM-2 contrast with the low but highly inducible expression levels of ICAM-1, and suggest important differences in their respective roles in pathophysiologic responses. ICAM-2 is the predominant LFA-1 ligand on unactivated endothelium, and therefore this pathway may be important in normal lymphocyte recirculation and memory T cell recruitment.[33] Because resting T cells express little or no ICAM-1, the LFA-1/ICAM-2 interaction may also be important in the interaction with antigen-presenting cells. In contrast, ICAM-1 is likely to be central to LFA-1-dependent adhesive functions during inflammatory and immune responses.

The LFA-1-dependent binding of cells to human endothelium and the LFA-1-dependent homotypic aggregation of JY cells can be blocked completely by combinations of MAbs to ICAM-1 and ICAM-2.[33] However, LFA-1-dependent aggregation of the T-lymphoblastoid cell line SKW3 cannot be inhibited by these MAbs, suggesting the existence of yet another LFA-1 ligand.[33] Using adhesion to immobilized LFA-1, it has been shown that indeed a third ligand is largely responsible for SKW3 phorbol ester-induced homotypic aggregation, and in fact several cell lines bind to LFA-1 in an ICAM-1- and ICAM-2-independent manner.[33] The structure of this ligand has not yet been defined, but functionally the ligand can be designated ICAM-3.

While the interaction of LFA-1 with ICAM-1 has been well established, the interaction of ICAM-1 with a second leukocyte integrin, Mac1 or CD11b/CD18, has been more controversial, with evidence both for and against such an interaction. A recent study using multiple cell binding assays, purified Mac1 and ICAM-1, and cell lines transfected with cDNAs for both ICAM-1 and Mac1, has shown conclusively that ICAM-1 does indeed interact with Mac1, both in purified form and on human endothelial cells.[35] The Mac1/ICAM-1 interaction is more temperature sensitive and lower in avidity than the LFA-1/ICAM-1 interaction and is blocked by a different series of ICAM-1-directed MAbs. Mutagenesis studies show in fact that Mac1 interacts with the third domain of ICAM-1, and thus ICAM-1 uses different Ig domains for adhesion to different leukocyte integrins.[25,35]

There are now several examples of receptor/ligand adhesion pairs that are implicated in signal transduction as well as cell–cell adhesion. For example, T cell activation via the CD2/LFA-3 interaction is well-established,[36] and it has recently been shown that ELAM-1 upregulates CD11/CD18 on adherent polymorphonuclear leukocytes (PMN).[37] Similarly, immobilized ICAM-1 can costimulate with MAb OKT3 T-cell activation and proliferation, under conditions where neither OKT3 alone nor ICAM-1 alone can stimulate.[15,38] These and other studies on the ICAM-1/LFA-1 interaction[39,40] underscore the active transient role played by adhesion molecules in regulating leukocyte adhesion to target cells, as might be expected when rapid and effective yet controlled responses to immune or inflammatory stimuli are required.

Numerous studies with blocking MAbs have now demonstrated the central importance of the CD11/CD18 integrins in leukocyte emigration *in vivo*,[4] although the role of ICAM-1 in these processes has been less well defined. Nevertheless, blocking MAbs to ICAM-1 are effective at inhibiting PMN emigration into phorbol ester-treated rabbit lung; at inhibiting eosinophil recruitment and airway responsiveness in a primate asthma model; at blocking primate renal allograft rejection; and at inhibiting rapid PMN adhesion to the rabbit vasculature in response to complement fragment C5a.[41-44] The *in vivo* role of this and other adhesion pathways is discussed in detail in Chapter 6.

VCAM-1/VLA-4 INTERACTION

When initial studies on endothelial–leukocyte adhesion were begun, two cytokine-induced endothelial cell adhesion molecules had been characterized and cloned, ELAM-1 and ICAM-1. ELAM-1, initially defined by a monoclonal antibody that partially inhibited PMN adhesion to IL-1-treated HUVECs,[45] is a member of the selectin or LECCAM family of adhesion molecules, which are structurally related to mammalian lectins, epidermal growth factor, and complement regulatory proteins.[3,7,46] Recent studies implicate sialylated fucosylated carbohydrate structures as components of the ligands for selectins, and confirm that this family of adhesion molecules are indeed mammalian lectins.[7,8]

Data from a number of laboratories show that neither ICAM-1 nor ELAM-1 fully account for the lymphocyte adhesion to cytokine-treated HUVECs observed *in vitro*. Although the ICAM-1/LFA-1 pathway clearly can mediate lymphocyte adhesion, blocking MAbs to LFA-1 can only weakly inhibit the binding of lymphocytic cells to activated endothelium.[12,14] Furthermore, in patients genetically deficient in CD18, and hence expressing no leukocyte LFA-1, lymphocyte recruitment into inflammatory sites is essentially normal, despite profound defects in recruitment of phagocytic cells,[15,47] indicating the existence of an ICAM/CD18-independent adhesion mechanism for lymphocytes.

Although ELAM-1 was initially characterized as an adhesion molecule for phagocytes, recent studies show that ELAM-1 can in fact support the adhesion of a subset of memory T cells, at least *in vitro*.[48-51] Thus, ELAM-1 may represent an LFA-1-independent pathway for the recruitment of certain T cell subsets *in vivo*. Nevertheless, lymphoblastoid cell lines do not bind to ELAM-1 expressed in COS cells,[10,52] while these cell lines do bind to cytokine-stimulated endothelial cells by an ICAM/LFA-1 independent pathway. For example, using an expressible ELAM-1 cDNA[53] and the anti-CD18 MAb 60.3,[54] the adhesion of various human leukemia cell lines to both HUVECs and ELAM-1-expressing COS cells were assessed, and their adhesion mechanisms were evaluated.[10] Some cell lines, such as the lymphoblastoid B-cell line JY, bound to HUVECs only through CD18, consistent with previously published data.[14] Others, such as the myelomonocytic cell lines HL60 and U937, bound to HUVECs partly through ELAM-1 and partly through some other pathway(s). Finally, most lymphocytic lines, such as

Jurkat (T cell) and Ramos (B cell), bound through neither CD18 nor ELAM-1, providing evidence that another adhesion pathway was operative on cytokine-treated HUVECs.[10]

With the existence of a new adhesion pathway for lymphocytic cell lines clearly established, cloning the inducible endothelial cell adhesion molecule responsible by direct expression techniques was attempted. To obviate the need for antibodies to detect expressing clones, a new screening procedure based on cell–cell adhesion itself was developed, which was used in conjunction with a subtraction technique to generate a sublibrary containing only cytokine-induced cDNAs.[10,53] Pools from the sublibrary were transfected into COS cells by spheroplast fusion. Cells were then screened for adhesion to fluorescently labelled Jurkat or Ramos cells 48 hours following fusion. By this means, a cDNA insert of 2.8 kb was isolated; when electroporated into COS cells, it resulted in the formation of large rosettes of either Jurkat or Ramos cells. The protein encoded by this clone was designated vascular cell adhesion molecule-1, or VCAM-1.[10]

The VCAM-1 cDNA insert contained a 1941 bp open reading frame, which encoded a predicted type I membrane protein, with a 24-amino-acid signal peptide, a 582-amino-acid extracellular domain, a 22-amino-acid transmembrane domain, and a short 19-amino-acid cytoplasmic tail. The amino acid sequence of VCAM-1 showed homologies to numerous proteins, all of the immunoglobulin superfamily. VCAM-1 was found to contain six Ig domains, all of the C2 or H type (see Figure 1–2). It can be placed within a subset of the Ig superfamily that contains other known adhesion molecules, including carcinoembryonic antigen (CEA), NCAM, and ICAM-1.[21–24,55,56]

Preliminary cell biologic studies on the interaction of VCAM-1 with lymphoblastoid cell lines suggested that its ligand might be an integrin. Although ICAM-1 interacts with leukocyte integrins, these had been eliminated as putative VCAM-1 ligands, and so other integrin subfamilies were examined. The β1 integrin family, also termed the *VLA proteins,* includes at least six receptors whose established interactions were with extracellular matrix proteins such as fibronectin, collagen, and laminin.[57] However, one of the β1 integrins, VLA-4 (see Figure 1–1), is restricted to lymphoid and myeloid cells,[57] which are VCAM-1-binding cells. In collaborative studies, the possible interaction of VLA-4 with VCAM-1 was investigated.[11] An anti-β1 antibody that blocks function of all the β1 integrins, as well as the specific anti-VLA-4 MAb HP2/1, blocked completely Ramos cell adhesion to either VCAM-1-expressing COS cells or to TNF-stimulated HUVECs, implying that VLA-4 is indeed responsible for VCAM-1-dependent adhesion. To confirm this indirect evidence, VLA-expressing stable cell lines were examined for binding to VCAM-1. Indeed, VLA-4-expressing transfectants, but not control VLA-2-expressing transfectants, bound to VCAM-1-expressing COS cells, confirming directly the VCAM-1/VLA-4 interaction.[11] Similar conclusions were reached independently using an established human CD18-deficient lymphoblastoid cell line and blocking MAbs.[58] The interaction of VCAM-1 with VLA-4 thus provided a second example of the interaction between the immunoglobulin and integrin superfamilies in cell–cell adhesion.

VLA-4 also mediates cell attachment to the CS1 site within the heparin II binding fragment of human plasma fibronectin.[59] The VLA-4-expressing transfectants described above, and Ramos cells, bind to a fibronectin fragment (Fn40) containing the CS1 site in a VLA-4-dependent manner,[11] and this interaction is completely inhibitable by soluble Fn40. In contrast, the interaction of VLA-4-expressing transfectants or Ramos cells with VCAM-1 cannot be inhibited by soluble Fn40. These results demonstrate that the VLA-4/VCAM-1 and VLA-4/Fn interactions are functionally distinct. It is possible that VLA-4-expressing mononuclear leukocytes may independently regulate these two adhesive functions during the inflammatory response *in vivo*. Thus, they may bind sequentially to VCAM-1 expressed on vascular endothelium and to fibronectin deposited at inflammatory sites.

Although VLA-4 is absent from PMN, recent studies from several laboratories indicate that VLA-4 is expressed on eosinophils from both normal and hypereosinophilic donors and that eosinophils can interact with VCAM-1, expressed on endothelium or CHO cells, or purified and immobilized on plastic.[60-63] Thus, the expression of VCAM-1 on the vascular endothelium may provide a selective means of eosinophil, as opposed to PMN, recruitment into inflammatory sites and may play a key role in eosinophil-related pathologies.

The ability of ICAM-1 to interact with more than one integrin raises the possibility that VCAM-1 may also possess other ligands, particularly in the light of the alternative splicing of VCAM-1 that is observed (see below). However, at present, VLA-4 appears to be the major, and perhaps the only, receptor for VCAM-1, because an absolute correlation between VLA-4 expression and ability to adhere to VCAM-1 has been found in over 20 cell types and cell lines examined to date.[64] However, it is likely that VLA-4 has other ligands. VLA-4-dependent homotypic aggregation has been described by several groups,[65,66] and this cell–cell aggregation cannot be accounted for by either VCAM-1 or fibronectin, strongly suggesting that at least one other VLA-4 ligand exists.[67]

The VCAM-1 cDNA originally isolated directs the expression of a functional adhesion molecule and encodes a protein containing six Ig homology units.[10] More recently, a longer form of VCAM-1 has been isolated (see Figure 1–2).[68] The IL-1-stimulated HUVEC library previously generated[10,53] was screened with radiolabeled VCAM-1 probe. Of three new clones isolated, two were different, as assessed by restriction digests. One of these clones, 1E11, was sequenced and was found to be identical in sequence to the original clone, except for an additional 276 nucleotides inserted between domains three and four of the original sequence. This insert is predicted to encode an additional Ig homologous domain, which has been designated domain 4, renumbering the following domains five through seven.[68]

The insert is preceded by the nucleotides AAG, common to splice junctions. Thus, the two forms likely represent alternately spliced products of the same precursor RNA, rather than the products of two different genes, a result confirmed by other studies.[69] The splice acceptor site of domain 4 is likely to be not optimal, resulting in skipping of this acceptor site in favor of the splice acceptor site of domain 5.

To discriminate between the two forms of VCAM-1 mRNA, the polymerase chain reaction (PCR) was used to amplify DNA fragments of different sizes representing the long and short forms.[68] In an examination of HUVECs isolated from umbilical cords of individual donors, it was found that both the long and short forms were induced in each of the three individuals tested, and that the longer form predominates. As other laboratories have begun to clone VCAM-1 from the published sequence, both forms have been detected, consistent with our data.[69,70]

Both forms support adhesion of VLA-4-expressing cells when expressed in COS7 cells,[68] indicating that domain 4 is not necessary for the adhesive function of VCAM-1. However, differences in affinity and/or specificity may occur, and a putative role for this domain in other as yet undefined aspects of VCAM-1 function is possible, in light of the tissue-specific alternative splicing observed with other adhesion molecules of the Ig superfamily, such as NCAM,[71,72] and the ability of ICAM-1 to adhere to two different leukocyte integrins. Two forms of VCAM-1 are expressed on the surface of rabbit vascular endothelial cells *in vitro,* and these may correspond to the two alternatively spliced forms of VCAM-1.[73] VCAM-1 is locally induced on rabbit arterial endothelium at sites of developing atherosclerosis, and alternative splicing may generate structures that have an important role in the pathogenesis of atherosclerosis.[69,73] Nevertheless, a MAb specific for the inserted domain of VCAM-1 (Benjamin et al., unpublished data), has recently been generated and this MAb stains VCAM-1 in all tissues so far examined, including inflamed endothelium and follicular dendritic cells in human tissues (unpublished observations). Thus, to date no evidence for preferential tissue expression has been obtained.

Having established that the longer seven domain form of VCAM-1 is predominant, both *in vitro* and *in vivo,* a recombinant soluble form of seven domain VCAM-1 has recently been generated.[74] A truncated cDNA lacking the transmembrane and cytoplasmic regions of VCAM-1 was constructed, stably expressed in Chinese hamster ovary (CHO) cells, and the secreted recombinant form of VCAM-1 (rsVCAM-1) was purified to homogeneity by immunoaffinity chromatography. Immobilized rsVCAM-1 is a functional adhesion protein and selectively binds only VLA-4-expressing cells, including human T and B lymphocytes, monocytes, NK cells, eosinophils, and certain lymphoblastoid cell lines.[63,74] T-cell subset analyses show that immobilized rsVCAM-1 shows a strong preference for memory T cells, with a further preference for CD8+ memory T cells.[74] In contrast, immobilized recombinant soluble ELAM-1 shows a preference under identical conditions for CD4+ memory T cells.[51] The preferential accumulation of the memory T cell subset in inflammatory lesions *in vivo*[75] has been attributed to their enhanced adhesive capacity for endothelium, as noted *in vitro*.[75,76] The preferential binding of cells of the memory phenotype to adhesion molecules such as ELAM-1[49–51] and VCAM-1[74] further emphasizes this possibility, while the propensity of ELAM-1 and VCAM-1 to bind preferentially CD4+ and CD8+ memory T cells, respectively, provides an added level of selectivity that may contribute to the pathophysiology of mononuclear leukocyte recruitment.

In recent studies, rsVCAM-1 and the anti-CD3 MAb OKT3 have been used to address the role of the VCAM-1/VLA-4 pathway in antigen-dependent T-cell activation.[77] Monocyte-depleted T cells proliferate upon exposure to coimmobilized OKT3 and rsVCAM-1 but to neither alone. In contrast, the anti-VLA-4 MAb HP1/2 fails to coactivate with OKT3, despite the fact that both rsVCAM-1 and HP1/2 support T cell adhesion at similar levels. Finally, immobilized rsELAM-1 failed to costimulate with OKT3. The data indicate that adhesive function alone is not sufficient for costimulatory activity and that the VCAM-1/VLA-4 pathway, like the ICAM/LFA-1 pathway, may play a role in regulating T-cell immune responses *in vivo*.[77]

Many laboratories have produced MAbs to cytokine-activated endothelial cells, and a number of these MAbs to previously unidentified endothelial cell antigens are now known to recognize VCAM-1, including MAbs E1/6, 4B9, 1E7, 2G7, and 1.4C3.[48,78–83] Studies with these MAbs and cytokine-stimulated endothelium *in vitro* confirm results from Northern blot analyses of VCAM-1 mRNA expression in TNF-treated HUVECs.[10] Cell surface VCAM-1 is detectable within two hours of cytokine stimulation, is maximal at 10 to 24 hours, and its expression is maintained for at least 72 hours in the continued presence of cytokine.[48,79–82]

The surface expression of ELAM-1, ICAM-1, and VCAM-1 on HUVECs *in vitro* is increased by the cytokines IL-1 and TNF and by bacterial lipopolysaccharide (LPS), and these three proinflammatory mediators show little or no selectivity in their ability to induce adhesion molecule expression, at least *in vitro*. However, recent studies suggest that the cytokine interleukin-4 (IL-4) may show such selectivity.[82–84] Masinovsky and coworkers found that markedly increased lymphocyte adhesion occurred to IL-4-treated macaque microvascular endothelial cells.[84] Moreover, strong synergism with IL-1 for lymphocyte adhesion was observed. A MAb, 6G10, which blocked up to 90% of lymphocyte adhesion to IL-4-treated cells was generated. Subsequently, 6G10 was found to bind to CHO cells expressing human VCAM-1. The results indicated that macaque VCAM-1 was largely responsible for the observed lymphocyte adhesion, and, more importantly, that VCAM-1 could be induced in endothelium by IL-4.[84] In parallel studies using HUVECs, Thornhill and coworkers found that IL-4 increased the adhesion of T cells, but not PMN, to endothelium by inducing an adhesion pathway that was ICAM/LFA-1 independent.[82,83] IL-4 was shown to induce an antigen defined by MAb 1.4C3, later identified as VCAM-1.[84]

The results suggest that IL-4 is unique in its ability to selectively modulate endothelial cell adhesion molecule expression. Its secretion by lymphocytes recruited to sites of inflammation or immune reactivity may serve to modulate local leukocyte accumulation, enhancing the transition from PMN-rich to mononuclear leukocyte-rich infiltrates during the evolution of inflammatory reactions. In addition, in light of the expression of VLA-4 by eosinophils,[60-63] it is intriguing to note that IL-4 expression *in vivo*, in either transplanted tumors or in transgenic animals, results in striking eosinophilic tissue infiltration.[85,86]

Using the MAbs described above, a number of immunohistochemical studies have been performed to evaluate the pathophysiologic tissue distribution of VCAM-1.[48,78,79,87-90] In general, little or no vascular staining of VCAM-1 has been found in normal tissues. In contrast, VCAM-1 is often expressed at inflammatory sites. E1/6 stains human vascular tissue in inflamed appendix, delayed hypersensitivity reactions, and insect bite.[78,79] 4B9 stains focally rheumatoid arthritic synovium.[88] More recent studies also show that VCAM-1 is expressed on the vascular tissue of rejecting human cardiac allografts, as assessed in endomyocardial biopsy tissues.[89,90] MAb E1/6 stains arterial as well as venular endothelium, marking VCAM-1 as distinct from ELAM-1, which is expressed exclusively in venular endothelium.[78,91]

Of particular note is a recent study characterizing the expression of rabbit VCAM-1, both *in vitro* and *in vivo*, using MAbs.[73] Importantly, in dietary hypercholesterolemic and Watanabe heritable hyperlipidemic rabbit models of atherosclerosis, VCAM-1 was found to be focally expressed by arterial endothelium overlying early foam cell-containing lesions. This lesion-localized expression suggests a central role for VCAM-1 in mononuclear leukocyte recruitment during atherogenesis.

Interestingly, VCAM-1, like ICAM-1, is not restricted to endothelial cells at inflammatory sites. Anti-VCAM-1 MAbs stain dendritic and macrophage-like cells in numerous normal and inflamed tissues, including skin, synovium, and lymph nodes.[79,87,88,90,92] In one report, functional assays were also performed, demonstrating the binding of VLA-4-expressing activated human B cells to VCAM-1-expressing human follicular dendritic cells.[87] The results suggest that the VCAM-1/VLA-4 pathway may play an important role in some aspects of the immune response, in addition to a role in mononuclear leukocyte recruitment through the vascular wall. As indicated above, the ability of immobilized rsVCAM-1 to costimulate with OKT3 and activate T cells[77] is consistent with such a possibility.

Recent studies indicate that the VCAM-1/VLA-4 pathway may also be important in lymphohemopoiesis.[93,94] Miyake and coworkers had obtained evidence that CD44 can function as an adhesion molecule on certain lymphoid cell lines by recognition of hyaluronate,[95,96] because MAbs to CD44 inhibit production of lymphoid and myeloid cells in long-term bone marrow cultures, suggesting a role for this adhesion pathway in hematopoiesis. However, they also found that adhesion of some lymphomyeloid cell lines to cloned stromal cells could not be explained on this basis.[96] Using CD44-independent adhesion of a murine lymphoid cell line to a murine stromal cell line as an assay, blocking MAbs were generated.[93] The first, PS/2, was found to recognize murine VLA-4, demonstrating a role for this integrin in lymphopoiesis within the bone marrow microenvironment.[93] The second set of MAbs, M-K/1 and M-K/2, define the counterreceptor on murine stromal cell clones derived from murine bone marrow.[93,94] The protein is similar in size to human VCAM-1, serves as an adhesion molecule on a murine endothelial cell line, and partial N-terminal sequence analysis generated a sequence highly homologous to human VCAM-1.[94] In collaborative studies, it has been confirmed that both MAbs recognize murine VCAM-1, because they bind to COS cells expressing a murine VCAM-1 cDNA and block adhesion of both murine and human lymphoblastoid cell lines to murine VCAM-1-expressing COS cells (Lobb et al., unpublished observations). In addition, the predicted N-terminal sequence of murine VCAM-1 from the cDNA matches that obtained from the protein isolated from the murine stromal cell line. The two MAbs to murine VCAM-1 selectively interfere with B lymphocyte formation when included in long-term bone marrow cultures, suggesting that the VCAM-1/VLA-4 pathway is important for B lymphocyte formation.[93,94]

SUMMARY

This review discusses two well-characterized pathways central to the adhesion of leukocytes to the vascular wall during inflammatory and immune responses, each pathway representing an Ig superfamily–integrin interaction. LFA-1-dependent cell adhesion involves at least three ligands, two of which have been characterized as members of the Ig superfamily, and ICAM-1-dependent cell adhesion involves at least two ligands, both of which are leukocyte integrins. VLA-4-dependent cell adhesion involves at least two ligands, one of which, VCAM-1, has been characterized as a member of the Ig superfamily. The two pathways share many similarities, but also show critical differences. LFA-1 is present on all leukocytes, and LFA-1-dependent adhesion is central to PMN adhesion and emigration. In contrast, VLA-4 is absent from PMN, and VLA-4-dependent pathways are likely to be more important in eosinophilic granulocyte recruitment. Both ICAM-1 and VCAM-1 can coactivate

T cells, suggesting a role for both pathways in the immune response as well as in leukocyte recruitment. Yet MAbs to ICAM/LFA-1 block mixed lymphocyte responses while those to VCAM/VLA-4 do not, and critical differences in the use of each pathway in normal and pathologic immune responses are certain to emerge as more data accumulate. Already, the VCAM-1/VLA-4 pathway appears to be important in lymphopoiesis, while the ICAM/LFA-1 pathway is not, as LAD patients deficient in CD18 exhibit essentially normal hematopoietic function. The data obtained to date suggest that further research on these two pathways should provide valuable insight into mechanisms of leukocyte recruitment into inflammatory and immune sites.

Acknowledgments: We thank Dr. T. Springer for access to information in press.

REFERENCES

1. Harlan, J.M. (1985) *Blood* 65, 513–525.
2. Stoolman, L.M. (1989) *Cell* 56, 907–910.
3. Osborn, L. (1990) *Cell* 62, 3–6.
4. Carlos, T.M., and Harlan, J.M. (1990) *Immuno. Rev.* 114, 1–24.
5. Springer, T.A. (1990) *Nature* 346, 425–434.
6. Pober, J.S. and Cotran, R.S. (1990) *Transplantation* 50, 537–544.
7. Brandley, B.K., Swiedler, S.J., and Robbins, P.W. (1990) *Cell* 63, 861–863.
8. Springer, T.A. and Lasky, L.A. (1991) *Nature* 349, 196–198.
9. Dustin, M.L., Staunton, D.E., and Springer, T.A. (1988) *Immunol. Today* 9, 213–215.
10. Osborn, L., Hession, C., Tizard, R., Vassallo, C., Luhowskyj, S., Chi-Rosso, G., and Lobb, R. (1989) *Cell* 59, 1203–1211.
11. Elices, M.J., Osborn, L., Takada, Y., Crouse, C., Luhowskyj, S., Hemler, M.E., and Lobb, R.R. (1990) *Cell* 60, 577–584.
12. Haskard, D., Cavender, D., Beatty, P., Springer, T., and Ziff, M. (1986) *J. Immunol.* 137, 2901–2906.
13. Martz, E. (1987) *Hum. Immunol.* 18, 3–37.
14. Dustin, M.L. and Springer, T.A. (1988) *J. Cell. Biol.* 107, 321–331.
15. Larson, R.S. and Springer, T.A. (1990) *Immunol. Rev.* 114, 181–217.
16. Rothlein, R., Dustin, M.L., Marlin, S.D., and Springer, T.A. (1986) *J. Immunol.* 137, 1270–1274.
17. Clark, E.A., Ledbetter, J.A., Holly, R.C., Dinndorf, P.A., and Shu, G. (1986) *Hum. Immunol.* 16, 100–113.
18. Marlin, S.D. and Springer, T.A. (1987) *Cell* 51, 813–819.
19. Dustin, M.L., Rothlein, R., Bhan, A.K., Dinarello, C.A., and Springer, T.A. (1986) *J. Immunol.* 137, 245–254.
20. Pober, J.S., Gimbrone, M.A., Jr., Lapierre, L.A., Mendrick, D.L., Fiers, W., Rothlein, R., and Springer, T.A. (1986) *J. Immunol.* 137, 1893–1896.
21. Simmons, D., Makgoba, M.W., and Seed, B. (1988) *Nature* 331, 624–627.
22. Staunton, D.E., Marlin, S.D., Stratowa, C., Dustin, M.L., and Springer, T.A. (1988) *Cell* 52, 925–933.
23. Williams, A.F. and Barclay, A.N. (1988) *Annu. Rev. Immunol.* 6, 381–405.
24. Hunkapillar, T. and Hood, L. (1989) *Adv. Immunol.* 44, 1–63.
25. Staunton, D.E., Dustin, M.L., Erickson, H.P., and Springer, T.A. (1990) *Cell* 61, 243–254.
26. Greve, J.M., Davis, G., Meyer, A.M., Forte, C.P., Yost, S.C., Marlor, C.W., Kamarck, M.E., and McClelland, A. (1989) *Cell* 56, 839–847.
27. Staunton, D.E., Merluzzi, V.J., Rothlein, R., Barton, R., Marlin, S.D., and Springer, T.A. (1989) *Cell* 56, 844–853.
28. Tomassini, J.E., Graham, D., DeWitt, C.M., Lineberger, D.W., Rodkey, J.A., and Colonno, R.J. (1989) *Proc. Natl. Acad. Sci. USA* 86, 4907–4911.
29. Marlin, S.D., Staunton, D.E., Springer, T.A., Stratowa, C., Sommergruber, W., and Merluzzi, V.J. (1990) *Nature* 344, 70–72.
30. White, J.M. and Littman D.R. (1989) *Cell* 56, 725–728.
31. Berendt, A.R., Simmons, D.L., Tansey, J., Newbold, C.I., and Marsh, K. (1989) *Nature* 341, 57–59.

32. Staunton, D.E., Dustin, M.L., and Springer, T.A. (1989) *Nature* 339, 61–64.
33. de Fougerolles, A.R., Stacker, S.A., Schwarting, R., and Springer, T.A. (1991) *J. Exp. Med.* 174, 253–267.
34. Nortamo, P., Salcedo, R., Timonen, T., Patarroyo, M., and Gahmberg, C.G. (1991) *J. Immunol.* 146, 2530–2535.
35. Diamond, M.S., Staunton, D.E., de Fougerolles, A.R., Stacker, S.A., Garcia-Aguilar, J., Hibbs, M.L., and Springer, T.A. (1990) *J. Cell Biol.* 111, 3129–3139.
36. Moingeon, P., Chang, H. C., Sayre, P.H., Clayton, L.K., Alcover, A., Gardner, P., and Reinherz, E.L. (1989) *Immunol. Rev.* 111, 111–144.
37. Lo, S.K., Lee, S., Ramos, R.A., Lobb, R., Rosa, M., Chi-Rosso, G., and Wright, S.D. (1991) *J. Exp. Med.* 173, 1493–1500.
38. Shimizu, Y., van Seventer, G.A., Horgan, K.J., and Shaw, S. (1990) *Immunol. Rev.* 114, 109–143.
39. Dustin, M.L. and Springer, T.A. (1989) *Nature* 341, 619–624.
40. Dransfield, I., Buckle, A.M., and Hogg, N. (1990) *Immunol. Rev.* 114, 29–44.
41. Barton, R.W., Rothlein, R., Ksiazek, J., and Kennedy, C. (1989) *J. Immunol.* 143, 1278–1282.
42. Wegner, C.D., Gundel, R.H., Reilly, P., Haynes, N., Letts, L.G., and Rothlein, R. (1990) *Science* 247, 456–459.
43. Cosimi, A.B., Conti, D., Delmonico, F.L., Preffer, F.I., Wee, S.L., Rothlein, R., Faanes, R., and Colvin, R.B. (1990) *J. Immunol.* 144, 4604–4612.
44. Argenbright, L.W., Letts, L.G., and Rothlein, R. (1991) *J. Leuk. Biol.* 49, 253–257.
45. Bevilacqua, M.P., Pober, J.S., Mendrick, D.L., Cotran, R.S., and Gimbrone, M.A., Jr. (1987) *Proc. Natl. Acad. Sci. USA* 84, 9238–9242.
46. Bevilacqua, M.P., Stengelin, S., Gimbrone, M.A., Jr., and Seed, B. (1989) *Science* 243, 1160–1165.
47. Anderson, D.C. and Springer, T.A. (1987) *Annu. Rev. Med.* 38, 175–194.
48. Graber, N., Gopal, T.V., Wilson, D., Beall, L.D., Polte, T., and Newman, W. (1990) *J. Immunol.* 145, 819–830.
49. Picker, L.J., Kishimoto, T.K., Smith, C.W., Warnock, R.A., and Butcher, E.C. (1991) *Nature* 349, 796–799.
50. Shimizu, Y., Shaw, S., Graber, N., Gopal, T.V., Horgan, K.J., Van Seventer, G.A., and Newman, W. (1991) *Nature* 349, 799–802.
51. Lobb, R.R., Chi-Rosso, G., Leone, D., Rosa, M., Bixler, S., Newman, B., Luhowskyj, S., Benjamin, C., Dougas, I., Goelz, S., Hession, C., and Pingchang Chow, E. (1991) *J. Immunol.* 147, 124–129.
52. Bevilacqua, M.P. and Gimbrone, M.A., Jr. (1989) in *Leukocyte Adhesion Molecules, Structure, Function, and Regulation.* (Springer, T.A., Anderson, D.C., Rosenthal, A.S., and Rothlein, R., eds.) pp. 215–223, Springer-Verlag, New York.
53. Hession, C., Osborn, L., Goff, D., Chi-Rosso, G., Vassallo, C., Pasek, M., Pittack, C., Tizard, R., Goelz, S., McCarthy, K., Hopple, S., and Lobb, R. (1990) *Proc. Natl. Acad. Sci. USA* 87, 1673–1677.
54. Beatty, P.G., Leadbetter, J.A., Martin, P.J., Price, T.H., and Hansen, J.A. (1983) *J. Immunol.* 131, 2913–2918.
55. Hemperly, J.J., Murray, B.A., Edelman, G.M., and Cunningham, B.A. (1986) *Proc. Natl. Acad. Sci. USA* 83, 3037–3041.

56. Oikawa, S., Nakazato, H., and Kosaki, G. (1987) *Biochem. Biophys. Res. Commun.* 142, 511–518.
57. Hemler, M.E. (1990) *Ann. Rev. Immunol.* 8, 365–400.
58. Schwartz, B.R., Wayner, E.A., Carlos, T.M., Ochs, H.D., and Harlan, J.M. (1990) *J. Clin. Invest.* 85, 2019–2022.
59. Wayner, E.A., Garcia-Pardo, A., Humphries, M.J., McDonald, J.A., and Carter, W.G. (1989) *J. Cell. Biol.* 109, 1321–1330.
60. Walsh, G.M., Mermod, J.J., Hartnell, A., Kay, A.B., and Wardlaw, A.J. (1991) *J. Immunol.* 146, 3419–3423.
61. Dobrina, A., Menegazzi, R., Carlos, T.M., Nardon, E., Cramer, R., Zacchi, T., Harlan, J.M., and Patriarca, P. (1991) *J. Clin. Invest.* 88, 20–26.
62. Bochner, B.S., Luscinskas, F.W., Gimbrone, M.A., Jr., Newman, W., Sterbinsky, S.A., Derse-Anthony, C.P., Klunk, D., and Schleimer, R.P. (1991) *J. Exp. Med.* 173, 1553–1556.
63. Weller, P.F., Rand, T.H., Goelz, S.E., Chi-Rosso, G., and Lobb, R.R. (1991) *Proc. Natl. Acad. Sci. USA* 88, 7430–7433.
64. Lobb, R., Hession, C., and Osborn, L. (1991) in *Cellular and Molecular Mechanisms of Inflammation: Vascular Adhesion Molecules* (Gimbrone, M.A., Jr. and Cochran, C.G., eds.) Academic Press, New York. In press.
65. Bednarczyk, J.L. and McIntyre, B.W. (1990) *J. Immunol.* 144, 777–784.
66. Campanero, M.R., Pulido, R., Ursa, M.A., Rodriguez-Moya, M., de Landazuri, M.O., and Sanchez-Madrid, F. (1990) *J. Cell Biol.* 110, 2157–2165.
67. Pulido, R., Elices, M.J., Campanero, M.R., Osborn, L., Schiffer, S., Garcia-Pardo, A., Lobb, R., Hemler, M.E., and Sanchez-Madrid, F. (1991) *J. Biol. Chem.* 266, 10241–10245.
68. Hession, C., Tizard, R., Vassallo, C., Schiffer, S.G., Goff, D., Moy, P., Chi-Rosso, G., Luhowskyj, S., Lobb, R., and Osborn, L. (1991) *J. Biol. Chem.* 266, 6682–6685.
69. Cybulsky, M.I., Fries, J.W.U., Williams, A.J., Sultan, P., Davis, V.M., Gimbrone, M.A., Jr., and Collins, T. (1991) *Am. J. Pathol.* 138, 815–820.
70. Polte, T., Newman, W., and Gopal, T.V. (1990) *Nucleic Acids Res.* 18, 5901.
71. Cunningham, B.A., Hemperly, J.J., Murray, B.A., Prediger, E.A., Brackenbury, R., and Edelman, G.M. (1987) *Science* 236, 799–806.
72. Dickson, G., Gower, H.J., Barton, C.H., Prentice, H.M., Elsom, V.L., Moore, S.E., Cox, R.D., Quinn, C., Putt, W., and Walsh, F.S. (1987) *Cell* 50, 1119–1130.
73. Cybulsky, M.I. and Gimbrone, M.A., Jr. (1991) *Science* 251, 788–791.
74. Lobb, R., Chi-Rosso, G., Leone, D., Rosa, M., Newman, B., Luhowskyj, S., Osborn, L., Schiffer, S., Benjamin, C., Dougas, I., Hession, C., and Chow, P. (1991) *Biochem. Biophys. Res. Commun.* 178, 1498–1504.
75. Pitzalis, C., Kingsley, G., Haskard, D., and Panayi, G. (1988) *Eur. J. Immunol.* 18, 1397–1404.
76. Damle, N.K., and Doyle, L.V. (1990) *J. Immunol.* 144, 1235–1240.
77. Burkly, L.C., Jakubowski, A., Newman, B.M., Rosa, M.D., Chi-Rosso, G., and Lobb, R.R. (1991) *Eur. J. Immun.* In press.
78. Rice, G.E. and Bevilacqua, M.P. (1989) *Science* 246, 1303–1305.
79. Rice, G.E., Munro, J.M., and Bevilacqua, M.P. (1990) *J. Exp. Med.* 171, 1369–1374.

80. Carlos, T.M., Schwartz, B.R., Kovach, N.L., Yee, E., Rosa, M., Osborn, L., Chi-Rosso, G., Newman, B., Lobb, R., and Harlan, J.M. (1990) *Blood* 76, 965–970.
81. Wellicome, S.M., Thornhill, M.H., Pitzalis, C., Thomas, D.S., Lanchberry, J.S.S., Panayi, G.S., and Haskard, D.O. (1990) *J. Immunol.* 144, 2558–2565.
82. Thornhill, M.H. and Haskard, D.O. (1990) *J. Immunol.* 145, 865–872.
83. Thornhill, M.H., Wellicome, S.M., Mahiouz, D.L., Lanchbury, J.S.S., Kyan-Aung, U., and Haskard, D.O. (1991) *J. Immunol.* 146, 592–598.
84. Masinovsky, B., Urdal, D., and Gallatin, W.M. (1990) *J. Immunol.* 145, 2886–2895.
85. Tepper, R.I., Pattengale, P.K., and Leder, P. (1989) *Cell* 57, 503–512.
86. Tepper, R.I., Levinson, D.A., Stanger, B.Z., Campos-Torres, J., Abbas, A.K., and Leder, P. (1990) *Cell* 62, 457–467.
87. Freedman, A.S., Munro, J.M., Rice, G.E., Bevilacqua, M.P., Morimoto, C., McIntyre, B.W., Rhynhart, K., Pober, J.S., and Nadler, L.M. (1990) *Science* 249, 1030–1033.
88. Koch, A.E., Burrows, J.C., Haines, K.G., Carlos, T., Harlan, J.M., and Leibovich, S.J. (1991) *Lab. Invest.* 64, 313–320.
89. Carlos, T., Gordon, D., Himes, V., Balassanian, E., Coday, A., and Fishbein, D. (1991) *Int. Soc. Heart Transpl. Abstr.* In press.
90. Briscoe, D.M., Schoen, F.J., Rice, G.E., Bevilacqua, M.P., Ganz, P., and Pober, J.S. (1991) *Transplantation* 51, 537–547.
91. Cotran, R.S. and Pober, J. (1988) in *Endothelial Cell Biology* (Simionescu, N. and Simionescu, M., eds.) pp. 335–347, Plenum, New York.
92. Griffiths, C.E.M., Barker, J.N.W.N., Kunkel, S., and Nickoloff, B.J. (1991) *Brit. J. Dermatol.* 124, 519–526.
93. Miyake, K., Weissman, I.L., Greenberger, J.S., and Kincade, P.W. (1991) *J. Exp. Med.* 173, 599–607.
94. Miyake, K., Medina, K., Ishihara, K., Kimoto, M., Auerbach, R., and Kincade, P.W. (1991) *J. Cell. Biol.* 114, 557–565.
95. Miyake, K., Medina, K.L., Hayashi, S.I., Ono, S., Hamaoka, T., and Kincade, P.W. (1990) *J. Exp. Med.* 171, 477–488.
96. Miyake, K., Underhill, C.B., Lesley, J., and Kincade, P.W. (1990) *J. Exp. Med.* 172, 69–75.

CHAPTER 2

Selectin/Carbohydrate-Mediated Adhesion of Leukocytes

James C. Paulson

Leukocyte recruitment to sites of tissue injury is mediated by multiple adhesion molecules, which are required for the attachment of the cells to the blood vessel endothelium, and their subsequent extravasation into the surrounding tissue. Leukocyte adhesion receptors are grouped into three main families: the integrin family, the super-immunoglobulin family, and the newly-described selectin or lectin-epidermal growth factor-complement related cell adhesion molecule (LECCAM) family, which is the focus of this chapter.[1] The two names commonly used for this family, selectin[2] and LECCAM,[3] both incorporate the term *lectin,* a carbohydrate-binding protein, reflecting the fact that this family of adhesion molecules recognizes carbohydrate ligands.[3-8]

Currently, there are three known members of the selectin family, endothelial leukocyte adhesion molecule-1 (ELAM-1) (LECCAM-2), CD62, (granule membrane protein-140 [GMP-140/ platelet activation dependent granule external membrane [PADGEM]/LECCAM-3), and Mel-14/LAM-1 (lymphocyte homing receptor/LECCAM-1/$_{gp}$90^{mel-14}/Leu8/TQ1/Ly-22). The ELAM-1 and CD62 selectins are differentially expressed on endothelial cells and platelets (CD62) in response to a variety of stimuli, and recruit neutrophils, other myeloid cells, and a subset of T lymphocytes to the sites of inflammation.[7,8] In contrast, Mel-14/LAM-1 is found on leukocytes. It was first described on lymphocytes, where it functions as a lymphocyte homing receptor to peripheral lymph nodes.[1,9,10] More recently, it has been found on neutrophils and other myeloid cells where it appears to participate in the recruitment of these cells to inflammatory sites.[11-16]

This chapter emphasizes the rapid advances in the understanding of the nature of the carbohydrate ligands of ELAM-1 and CD62 and emerging views on the roles of the selectins in the recruitment of leukocytes to sites of inflammation and tissue injury. For additional information on Mel-14/LAM-1 and its role in lymphocyte recirculation, the reader is referred to Chapter 3 and to several other recent reviews.[9,10]

THE SELECTIN FAMILY—A HOMOLOGOUS GENE FAMILY

Within a remarkably brief period, cDNA sequences for the three selectins were reported from multiple groups, revealing a new homologous family of leukocyte adhesion molecules.[17-27] As shown in Figure 2-1, each of the selectins has an NH_2-terminal lectin domain (~120 residues), an epidermal growth factor (EGF) domain (~30 residues), 2–9 complement regulatory (CR) repeat sequences (62 residues each), a transmembrane domain, and a short COOH-terminal cytoplasmic domain.

The Lectin Domains

The lectin domain was recognized to be homologous to the family of C-type

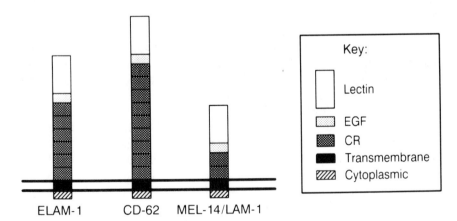

Figure 2-1. Domain structures of the selectins.

lectins described by Drickamer,[28,29] who had identified spacially dispersed but conserved residues in a variety of mammalian carbohydrate binding proteins whose binding activities, like the three selectins, were calcium-dependent. While the homology between the lectin domains of the selectins and other known mammalian lectins ranges between 25% and 30%, the homology among the lectin domains of the selectin family is approximately 60%.[17,30]

Identification of the lectin domains in this new family of cell adhesion molecules drew wide attention to the potential for carbohydrate ligands in mediating leukocyte–endothelial adhesion. Although Rosen and colleagues had earlier provided evidence that adhesion properties of Mel-14/LAM-1 were carbohydrate-mediated,[31–34] the carbohydrate binding properties of the other two selectins were not previously recognized.[17] After identification of the lectin domains in ELAM-1 and CD62, their carbohydrate-binding properties were rapidly demonstrated, and it is now generally acknowledged that all three selectins mediate leukocyte–endothelial adhesion through recognition of carbohydrate ligands.[3–7]

CR and EGF Domains

Less is known about the functions of the CR repeat and the EGF domains of the selectins. The CR repeats bear homology to 62 residue repeats found in several complement pathway proteins, (e.g., CR_2) which bind complement proteins C3b or C4b.[17,23,24] However, this homology has not yet provided a clue to the function of the CR domain.

In addressing the function of the EGF domain, Siegelman et al.[24,26] have shown that an antibody (anti-Ly-22) to the EGF domain of the Mel-14/LAM-1 blocks adhesion of lymphocytes to high endothelial venules (HEV). They have proposed that the EGF domain may be directly involved in ligand recognition, perhaps by recognition of a protein determinant. Similar results and interpretations have been reported by Kansas et al.[10] using an antibody whose epitope was mapped to the EGF domain. This view implies the possibility that the physiologically relevant ligand of Mel-14/LAM-1, and perhaps the other two selectins, is a specific protein that carries both carbohydrate and protein determinants[35] (see also Chapter 3).

Soluble Forms of CD62

Johnston et al. identified alternatively spliced cDNA clones of CD62 that had a deletion of the sequence coding for the transmembrane domain, predicting that soluble forms of CD62 might be produced. Gamble et al.[36] have demonstrated that soluble forms of CD62 can inhibit the low levels of adhesion of activated neutrophils to resting endothelial cells *in vitro*, suggesting a potential role of soluble CD62 in limiting inflammatory reactions to the site of activated endothelium. The mechanism of this inhibition has not been elucidated but is likely to be due to the adhesion of CD62 to the carbohydrate ligands on glycoproteins of the activated neutrophils that mediate adhesion to resting endothelial cells (e.g., the LFA-1 and Mac-1 integrins that contain SLe^x-related carbohydrate groups.[37,38]) The potential significance of these observations rests on the presence of soluble forms of CD62 *in vivo,* which have not yet been reported.

CELLULAR DISTRIBUTION AND EXPRESSION OF THE SELECTINS

All three selectins are involved in the recruitment of neutrophils and other myeloid cells to sites of tissue injury, but there are fundamental differences in their distribution, activation, and mode of expression.

Mel-14/LAM-1 is constitutively expressed on some lymphocytes, neutrophils, monocytes, and other myeloid cells.[11,14,27,39,40] The role of Mel-14/LAM-1 in the recirculation of lymphocytes to peripheral lymph nodes is well-documented, and there is ample evidence that this selectin can mediate adhesion by recognition of carbohydrate ligands on the endothelial cells of high endothelial venules (see Chapter 3). Mel-14/LAM-1 has also been demonstrated to play a role in the recruitment of neutrophils to inflammatory sites *in vivo.*[12,15,16,40] However, there is as yet no direct evidence for carbohydrate ligands of Mel-14/LAM-1 on the endothelial cells of postcapillary venules in inflamed tissue.

ELAM-1 and CD62 are expressed on endothelial cells and on platelets and endothelial cells, respectively. *In vitro,* ELAM-1 mediates adhesion of neutrophils, monocytes, eosinophils, a subset of lymphocytes, and certain carcinoma cells (Table 2-1). CD62 binds similar cell types, although a detailed direct comparison of the cell type specificities for ELAM-1 and CD62 has not yet emerged (see Table 2-1).

Expression of ELAM-1 and CD62 is inducible, and neither is expressed on unactivated endothelial cells or platelets. Yet, they differ in their mechanisms of expression and the different activators that induce their expression (Table 2-2). ELAM-1 is expressed only on endothelial cells following induction by cytokines such as interleukin-1β (IL-1β), tumor necrosis factor-α (TNF-α), and lymphotoxin, and also by bacterial endotoxin, interferon-γ, and the neuropeptide substance P.[17,41–43] In endothelial cell culture *in vitro,* these substances activate the transcription of ELAM-1, resulting in peak cell surface expression at four to six hours, decreasing to basal levels by 24 hours.[17,41] Although there is less information about the kinetics of expression *in vivo,* available evidence suggests that the time course is similar.[44] Although ELAM-1 expression is typically transient, it appears to be chronically expressed in certain inflammatory conditions, such as psoriasis and rheumatoid arthritis.[45–47]

Table 2-1. Cell Types Bound by ELAM-1 and CD62

Cell Type	ELAM-1	CD62	References
Leukocytes			
Neutrophils	Yes	Yes	Bevilacqua et al.[41]; Geng et al.[2]; Larsen et al.[48]; Toothill et al.[50]; Hamburger and McEver[120]
Monocytes	Yes	Yes	Walz et al.[59]; Larsen et al.[48]; Lobb et al.[97]
Eosinophils	Yes	??	Kyan-Aung et al.[98]
NK cells	Yes	??	Goetz et al.[58]; Shimizu et al.[99]
T cells	Yes (subset)	??	Picker et al.[45]; Lobb et al.[97]
B cells	No	??	
Erythrocytes	No	No	
Platelets	No	No	
Tumor Cells			
Carcinoma cells	Yes	Yes	Rice and Bevilacqua[121]; Kishimoto et al.[113]; Walz et al.[59]; Polley et al.[55]
Melanoma cells	No	??	Rice and Bevilacqua[121]

CD62 expression does not require *de novo* synthesis because it is stored in secretory granules (or Weibel-Palade bodies) in both platelets and endothelial cells. Thus, within minutes of activation of either cell type by thrombin, histamine, or phorbol esters, CD62 is rapidly redistributed to the surface of the cell where it can adhere neutrophils, monocytes, and other cells.[2,48–51]

Patel et al.[52] have found that endothelial cells also express CD62 in response to low levels of hydrogen peroxide or other oxidizing agents through the production of free radicals. While endothelial cells normally reinternalize CD62 within minutes of activation, induction by free radicals produced

Table 2-2. Biological Properties of ELAM-1 and CD62

Property	ELAM-1	CD62
Distribution	Endothelial cells	Endothelial cells Platelets
Activators	Interleukin 1-β Tumor necrosis factor Lymphotoxin Interferon - γ Substance P Bacterial endotoxin (LPS)	Thrombin Histamine H_2O_2
Duration of expression	2 to 24 hours	1 to 15 Minutes

prolonged expression of the selectin.[52,53] The prolonged expression appears to result from lack of endocytosis of surface CD62, resulting in a higher level of CD62-mediated adhesiveness than obtained with other activating agents. Because neutrophils release oxidizing agents and free radicals following activation, initial recruitment of neutrophils by transiently expressed CD62 could effectively prolong the expression of CD62 through free radical generation by the bound neutrophils.[52]

CARBOHYDRATE LIGANDS OF THE SELECTINS

There has been rapid progress in identification of the discrete carbohydrate structures recognized by ELAM-1 and CD62 in mediating leukocyte–endothelial cell adhesion.[3-7] The carbohydrate determinants recognized by the selectins are typically found as terminal structures on carbohydrate groups of one or more glycoproteins and/or glycolipids. In this respect, carbohydrate ligands differ from protein ligands in that multiple cell surface components have similar if not identical carbohydrate structures produced by the "set" post-translational glycosylation machinery of that cell. It is possible that a single cell surface glycoprotein is the physiologically relevant receptor of a selectin, providing that only one glycoprotein carries the carbohydrate ligand (protein-specific glycosylation), or that a selectin recognizes protein epitopes in addition to carbohydrate epitopes. But it is equally likely that the selectins recognize multiple cell surface molecules as receptors by virtue of their carrying common carbohydrate ligands.

Mel-14/LAM-1

Although the relevant carbohydrate ligand(s) recognized by Mel-14/LAM-1 has not yet been reported (see Chapter 3), various carbohydrates have been demonstrated to inhibit Mel-14/LAM-1 adhesion, including mannose-6-phosphate, PPME (a yeast polymannan containing the mannose-6-phosphate determinant), and a variety of anionic polysaccharides such as fucoidan and heparin.[31-34] The interaction with PPME is sufficiently avid that a PPME affinity matrix was successfully used to purify Mel-14.[23] Although PPME and the other carbohydrate inhibitors provide potential clues to the natural ligands, they are not found as natural components of cell surface carbohydrate groups.

Perhaps most relevant to understanding the nature of the physiologic ligand of Mel-14/LAM-1 was the observation of Rosen et al.[32] that sialidase treatment of peripheral lymph node sections abolished lymphocyte adhesion to high endothelial venules in the Stamper-Woodruff assay. Intravenous injection of sialidase in mice also abolished lymphocyte recirculation to peripheral lymph nodes, suggesting that inhibition of lymphocyte adhesion *in vitro* is relevant to the adhesion event that mediates lymphocyte homing.[34] Together, these results suggested that sialic acid is an essential component of the carbohydrate ligand of Mel-14/LAM-1 on high endothelial venules of peripheral lymph nodes. This observation is of particular interest in view of the recent evidence indicating that the other two members of the selectin family, ELAM-1 and CD62, also require sialic acids as essential components of their carbohydrate ligands.[5,6,55] In a recent report, Imai et al.[35] used a Mel-14-Ig chimera to identify a 50 kd glycoprotein in peripheral lymph nodes that exhibits the properties of the Mel-14/LAM-1 ligand, including sialidase-sensitive binding. Thus, in lymph nodes, the effective ligand of Mel-14/LAM-1 could be a single glycoprotein.

ELAM-1

In a rapid series of reports from several laboratories, the carbohydrate ligand of ELAM-1 was demonstrated to be a member of a class of sialylated and fucosylated structures related to the sialylated Lewis-X blood group antigen (SLe^x) (Diagram 1).[56-60]

Le^x and SLe^x were previously documented to be present as terminal structures on cell surface glycoproteins and glycolipids of neutrophils and promyelocytic cell lines,[38,61-66] and Le^x had even been recognized earlier as a differentiation marker, CD15, on neutrophils and monocytes.[63,67-70] However, until the discovery of the selectin family of adhesion molecules, the functional significance of these structures in cell adhesion was not appreciated.

SLe^x as a Ligand for ELAM-1

A variety of approaches used by several laboratories has provided evidence that the SLe^x structure is a ligand for ELAM-1. Lowe et al.[57] demonstrated that transfection of a cDNA for the Lewis blood group fucosyltransferase

$$Gal\beta1,4 \diagdown$$
$$\quad\quad\quad\quad GlcNAc\text{-}R \quad\quad Lewis\ X\ (Le^x)$$
$$Fuc\alpha1,3 \diagup$$

$$SA\alpha2,3Gal\beta1,4 \diagdown$$
$$\quad\quad\quad\quad\quad\quad GlcNAc\text{-}R \quad\quad Sialyl\text{-}Lewis\ X\ (SLe^x)$$
$$Fuc\alpha1,3 \diagup$$

Diagram 1. Structures of Lewis X and Sialyl-Lewis X.

(Galβ1,3/4GlcNAc α1,3 fucosyltransferase) into Chinese hamster ovary (CHO) cells resulted in the expression of the Le^x and SLe^x antigens and the simultaneous ability of the transfected cells to adhere to ELAM-1 on TNF-α-activated human umbilical vein endothelial cells (HUVECs). Sialidase treatment of the cells abolished their ability to adhere to activated HUVECs, indicating that a sialylated structure was required for adhesion. Further evidence suggesting SLe^x as the ligand was the observation that an pro-myelocytic leukemia-60 (HL-60) cell clone that expressed SLe^x bound to ELAM-1, while another clone that failed to express SLe^x did not.

Similar results were obtained by Phillips et al.[56] using CHO glycosylation mutants, which unlike the wild-type cells, expressed fucosyltransferase activities that synthesized both Le^x and SLe^x (LEC11) or Le^x only (LEC12) as terminal sugar structures on cell surface glycoproteins.[71-73] Only LEC11 cells bound to ELAM-1 on activated HUVECs, and the adhesion was abolished by pretreatment of the LEC11 cells with sialidase, implicating SLe^x as the ligand.

Inhibition of ELAM-1-mediated adhesion by anti-SLe^x antibodies, but not anti-Le^x or other carbohydrate-specific antibodies tested, provided additional evidence for SLe^x as a carbohydrate ligand.[56,59] Walz et al.[59] also showed that glycoproteins reported to carry the SLe^x determinant could also inhibit adhesion (human amniotic mucin) or support adhesion when adsorbed to plastic (fucosylated α$_1$-acid glycoprotein). In a related approach, Phillips et al.[56] and Polley et al.[55] used glycolipids incorporated into liposomes as inhibitors. Liposomes containing SLe^x as a terminal structure inhibited adhesion of HL-60 cells and neutrophils to ELAM-1 on activated HUVECs, while glycolipid liposomes with related structures, missing either the sialic acid or fucose, exhibited little or no inhibition.

Vim-2 Antigen as a Ligand

A different but related carbohydrate structure recognized by the Vim-2 antibody has also been proposed as a carbohydrate ligand of ELAM-1 by Tiemeyer et al.[60] Vim-2 antigen was deduced to be a ligand for ELAM-1 using purification of neutrophil glycolipids and determining the structure of the purified glycolipids that mediated adhesion of ELAM-1 transfected COS cells. The relationship of the structure of the Vim-2 antigen (CD65) to SLe^x is illustrated in Figure 2-2. While both are sialylated and fucosylated structures, the Vim-2 antigen has the fucose separated from the sialic acid by two additional sugar residues in a polylactosamine core. The studies of Phillips et al.[56] examining inhibition of ELAM-1-mediated adhesion using glycolipid liposomes employed a glycolipid (S-diLe^x) incorporating features of both SLe^x and Vim-2 antigen (see Figure 2-2), shedding little light on the relative efficacy of the two structures as a ligand. More recently, Polley et al.[55] have found that glycolipid liposomes containing either the S-diLe^x or

SLe^x structures (see Figure 2-2) are both potent inhibitors of ELAM-1-mediated cell adhesion.

Walz et al.[59] and Lowe et al.[74] have investigated the Vim-2 antigen as a potential ligand for ELAM-1 and were unable to confirm its role in mediating cell adhesion. Vim-2 antibodies were unable to block adhesion, and cells containing the Vim-2 antigen but not the SLe^x structure were unable to bind to recombinant ELAM-1 or to ELAM-1 expressed on TNF-α-activated endothelial cells. The reason for the apparently conflicting results of these groups and those of Tiemeyer et al.[60] is not clear at present. It is possible that the cells examined by Walz et al.[59] and Lowe et al.[74] may not have expressed sufficient Vim-2 antigen to support adhesion by ELAM-1. Another possibility addresses the rigorous but technically difficult approach used by Tiemeyer et al.[60] to characterize the neutrophil glycolipids recognized by ELAM-1. Glycolipids were purified by several chromatography steps to single bands on thin layer chromatography. However, it is difficult to separate related isomers (that may be bioactive) by these techniques, particularly when the sugar chains are 8 to 10 residues in length.[60] Further clarification of the

Type	Terminal Sequence	Structure	ELAM-1	CD-62
Soluble sugar	Le^x		−	+
Soluble sugar	SLe^x		−	+
Glycolipid	SLe^x		+	+
Glycolipid	S-diLe^x		+	+
Glycolipid	Vim-2 antigen		+	?

Key: ◀ Sialic Acid ■ GlcNAc ◆ Fuc □ Glc ○ Gal ⊏ Ceramide

Figure 2-2. Comparison of carbohydrate ligands of ELAM-1 and CD62. Structures of carbohydrate ligands shown to interact with ELAM-1 and/or CD62 are shown in symbol form. The sequence of the largest structure (S-diLe^x) is NeuAcα2,3-Galβ1,4(Fucα1,3)GlcNAcβ1,3Galβ1,4(Fucα1,3)GlcNAcβ1,3Galβ1,4Glcβ ceramide. All the other structures are truncated versions of this structure, with linkages between the remaining sugars remaining the same.

relative affinities of the SLe^x and Vim-2 antigen structures for ELAM-1 and their role in mediating adhesion of leukocytes *in vivo* awaits a more detailed study of the fine specificity of ELAM-1 for its carbohydrate ligand(s).

CD62

Evidence from several laboratories suggests that CD62, like ELAM-1, recognizes SLe^x as a high-affinity ligand. Larsen et al.[48] demonstrated that anti-Le^x antibodies (anti-CD15) could inhibit CD62-mediated adhesion of neutrophils. In addition, a human milk oligosaccharide containing the terminal Le^x structure (LNF III) exhibited sufficient affinity for CD62 to inhibit adhesion, while a related isomer (LNF I) did not. However, Corral et al.[75] and Moore et al.[76] found that CD62-mediated adhesion of neutrophils could be inhibited by prior treatment of the cells with sialidase to remove sialic acids. Because sialidase treatment actually increases cell surface Le^x (by converting SLe^x to Le^x), Le^x cannot account for the CD62-mediated adhesion of these cells.

Polley et al.[55] using an assay similar to that used by Larsen et al.[48] demonstrated that an SLe^x-containing oligosaccharide was a 30-fold more potent inhibitor of CD62-mediated adhesion than Le^x (LNF III). Furthermore, a CHO-cell glycosylation mutant containing SLe^x (LEC11) bound to CD62, while a mutant with only Le^x did not. The results suggested that soluble oligosaccharides containing either Le^x and SLe^x can inhibit adhesion *in vitro* at sufficiently high concentrations, but only SLe^x binds with high enough affinity to serve as a cell surface ligand that supports cell adhesion.

Although both ELAM-1 and CD62 appear to recognize SLe^x as a high-affinity ligand, it is premature to conclude that they have an identical ligand specificity. Indeed, the specificity of these two selectins may differ in their ability to distinguish between the various carbohydrate groups that carry the terminal SLe^x carbohydrate structure (see below), may include recognition of protein epitopes, and may even have a broader specificity that extends beyond SLe^x-related structures. Thus, it should be anticipated that differences in fine specificity may ultimately be found to influence biologic properties of the two selectins, such as their cell type selectivities.

DISTRIBUTION AND BIOSYNTHESIS OF SLe^x

Identification of Le^x and SLe^x as Leukocyte Antigens

Le^x and SLe^x were originally described as structures related to the Lewis blood group antigens.[77] Interest in the Le^x structure intensified when it was shown to represent a stage-specific antigen (SSEA-1) in murine embryogenesis.[78,79] Le^x was soon thereafter recognized as a lineage-specific antigen found on mature granulocytes, monocytes, and natural killer (NK) cells, but not on erythrocytes or mature T and B lymphocytes.[6,80] Sialylated Le^x antigens were also reported to have a similar restricted distribution on mature hemopoietic cells, although both the Le^x and SLe^x antigens were found to be more broadly distributed in leukemic cells and immature cells of both the erythroid and lymphoid lineages.[63,80,81] In addition to leukocytes, the Le^x and SLe^x antigens have been detected on a variety of normal human tissues, particularly in the gut and kidney,[82] and are expressed in a high percentage of lung and colon carcinomas.[83–85]

Expression of SLe^x on Glycoproteins and Glycolipids of Leukocytes

Direct analysis of cell surface glycoconjugates on granulocytic leukemia cells, promyelocytic cells, mixed leukocytes (buffy coats), and human neutrophils has revealed that both glycoproteins[37,38,61,64,86] and glycolipids[66] contain carbohydrate groups with terminal Le^x and SLe^x.

The most detailed structure analysis has been done on the N-linked carbohydrate groups of glycoproteins.[38,61,64,86] Although an exhaustive listing of these complex structures (12–24 sugar residues) is not warranted here, several aspects of structure revealed by these studies may be relevant to understanding the interactions of selectins with their carbohydrate ligands. There are literally dozens of different N-linked oligosaccharides that carry terminal SLe^x due to the branching of N-linked carbohydrate groups, extension of individual branches by repeating N-acetyl-lactosamine sequences (Galβ1,4 GlcNAc), and microheterogeneity (e.g., ±Fucose). The shortest chains carry the SLe^x structure attached directly to the Man$_3$GlcNAc$_2$ core structure while the longest carries SLe^x attached to branches with 2 to 3 N-acetyllactosamine (polylactosamine) repeats.[38,64,86] Thus, the SLe^x substituent would variably extend 7 to 13 residues from the polypeptide chain. Assuming that each sugar unit is approximately 6 Å, the SLe^x structure on glycoproteins would extend 42 to 78 Å from the polypeptide if fully extended.

Several specific leukocyte proteins have been demonstrated to carry N-linked carbohydrate groups with the SLe^x structure, including the lysosomal LAMP-1 and LAMP-2 glycoproteins[86] and the CD18 integrins.[37,38] Using anti-CD15 antibodies, Skubitz and Snook[37] showed that Le^x determinants were carried on a variety of cell surface adhesion molecules including LFA-1, Mac-1, gp150/95, and CR1. Direct analysis of the N-linked oligosaccharides of affinity-purified leukocyte CD18 integrins (LFA-1, Mac-1, and gp150/95) by Asada et al.[38] demonstrated the presence of sialylated Le^x structures. Because all glycoproteins are glycosylated through a common pathway, other surface glycoproteins of leukocytes can also be expected to carry N-linked oligosaccharides with the SLe^x structure.

Although less work has been done on the direct structure analysis of leukocyte glycolipids,[60,66,87] the structural variation can be expected to be similar to the structures represented in Figure 2-2, including longer structures with additional N-acetyl-lactosamine repeat units ending in the same terminal structures (SLe^x, S-diLe^x, and Vim-2 antigen).

The information available to date raises important questions concerning the "effective" leukocyte ligands for ELAM-1 and CD62. *In vitro* both glycoproteins and glycolipids containing SLe^x interact with the selectins.[56,59] However, it is possible that one or more of the cell surface glycoproteins represents the predominant ligand(s) for ELAM-1 and CD62. Thus, it will ultimately be important to understand the degree to which the fine specificity of the selectins for their carbohydrate ligands, and the physical accessibility of the carbohydrate groups, determine which cell surface molecules are the "effective ligands" on the surface of the leukocytes.

Biosynthesis of SLe^x

Formed as a terminal structure on the carbohydrate groups of glycoproteins and glycolipids, SLe^x is synthesized by the sequential addition of sialic acid and fucose to the core N-acetyllactosamine sequence. These additions are among the last steps in the synthesis of the carbohydrate group. Although the sequence of reactions could, in principle, involve fucosylation first, it is generally accepted that the order of reactions requires sialylation first, because the α2,3 sialyltransferase cannot attach sialic acid to fucosylated substrates (Diagram 2).[71,83,88]

An added complexity to this fairly simple biosynthetic pathway is the existence of at least three α1,3 fucosyltransferases, which are capable of forming the Le^x determinant.[71,89–94] Based on easily distinguishable substrate specificities for all three enzymes, Mollicone et al.[89] identified a "Lewis blood group type," a "liver (or plasma) type," and a "myeloid type." The Lewis blood group type is unique in its ability to transfer fucose to either the 3 or 4 position of GlcNAc depending on whether the substrate presented has the Galβ1,4GlcNAc or Galβ1,3GlcNAc linkages, respectively.[89,91,93,94] Both the Lewis type and liver type transfer to sialylated substrates (NeuAc2α,3Galβ1,4GlcNAc) with good efficiency to yield the SLe^x sequence. In contrast, the myeloid type transfers poorly to sialylated substrates, if at all. Although Mollicone et al.[89] were able to identify tissues and secretions that expressed predominantly one of three enzymes, most sources express two or more of the enzymes,[89,92,94] making absolute determinations of substrate specificity difficult.

Two of the fucosyltransferases have been cloned and examined for their enzymatic specificities by expression in CHO cells or COS cells.[57,58,73,74,95] The Lewis blood group fucosyltransferase cloned by Kukowska-Latallo et

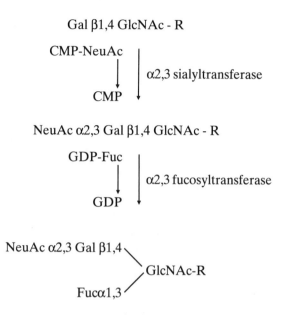

Diagram 2. Biosynthesis of SLe^x.

al.[95] has been demonstrated to produce both the SLe^x and Le^x antigens on the surface of transfected CHO cells, allowing them to adhere to ELAM-1 expressed on activated HUVEC.[74] Conflicting results have been obtained with the cloned "myeloid type" fucosyltransferase. Lowe et al.[57,74] found that transfected CHO cells produced the Le^x and Vim-2 antigens, but did not produce the SLe^x antigen, and were not able to adhere to ELAM-1 on activated endothelial cells. In contrast, Goelz et al.[58] found that the same cDNA, cloned independently, caused transfected COS cells to adhere to recombinant ELAM-1-coated beads and plastic plates. The discrepancy between the two results is not clear at present. The higher levels of fucosyltransferase activity expressed in the latter case[58,74] may have allowed a low level of SLe^x to be expressed. Another possibility is that this enzyme produces an alternative ligand for ELAM-1 in transfected COS cells that is not produced in transfected CHO cells. Equally puzzling, however, is that this "myeloid type" fucosyltransferase does not appear to account for the abundant expression of SLe^x on granulocytes and promyelocytic cell lines. These results suggested that another fucosyltransferase in myeloid cells (expressed at a lower level) may be responsible for the synthesis of SLe^x or that another factor in myeloid cells modulates the acceptor specificity of the enzyme.[74]

Relevance of SLe^x Expression to the Cell Selectivities of ELAM-1 and CD62

Despite the extensive literature cited above, there is limited information to evaluate the extent to which the SLe^x ligand determines the full range of leukocyte selectivities of ELAM-1 and CD62 *in vivo*. A summary of the blood cell types reported to bind to ELAM-1 and CD62 are listed in Table 2-1. The list is not complete, as all the cell types reported to bind to ELAM-1 have not been systematically tested for their adhesion to CD62. Because both selectins recognize SLe^x as a ligand, the two lists would be identical if recognition of this carbohydrate was the sole factor in determining cell type specificity. However, much of the detailed information obtained on the expression of SLe^x on cell surface carbohydrate groups has been obtained using leukemic cells that may not reflect the expression on circulating leukocytes.[80] Thus, the hypothesis that SLe^x mediates the full spectrum of cell selectivities of the two selectins has not been thoroughly addressed.

Several reports have examined both SLe^x expression and ELAM-1 mediated adhesion of cells and have documented a good correlation.[56,57,59] For the most part, these reports have examined promyelocytic (e.g., HL-60) or monocytic (e.g., THP-1) cell lines and not blood leukocytes. However, SLe^x is well-documented to be present on blood neutrophils, and sialylated forms of Le^x have been reported on circulating monocytes,[96] suggesting that SLe^x is available as a cellular ligand on these cells. Similarly, NK cells have been reported to adhere to ELAM-1,[58,97] and in a separate study have been demonstrated to contain the SLe^x antigen.[80]

The situation for eosinophil and T-cell adhesion to ELAM-1 is less clear. Kyan-Aung et al.[98] have demonstrated ELAM-1-mediated adhesion of eosinophils to activated HUVECs. Although expression of SLe^x on these cells has not been examined directly, Macher and Beckstead[87] did not detect expression of the biosynthetic precursor, Le^x, which would be expected to be found if SLe^x were present. A subclass of T cells associated with high ELAM-1 expression on postcapillary venules in chronic skin inflammation has been demonstrated to adhere to recombinant ELAM-1.[45,99] The adherent cells represent 10%–15% of the total population of peripheral T lymphocytes, express the HECA-452 (CLA) antigen, and are enriched in the T-memory phenotype.[45,99] As discussed above, immature lymphoid precursors and a high percentage of lymphomas are found to express SLe^x, but Ohmori et al.[80] were unable to detect SLe^x on peripheral blood lymphocytes. The results imply that SLe^x may not be available as a ligand on mature T cells. For both eosinophils and T cells, it is possible that alternative ligands to SLe^x mediate adhesion to ELAM-1, although definitive studies to address this point have not been forthcoming.

Despite rapid advances in understanding carbohydrate ligands of ELAM-1 and CD62, the current picture regarding their ligands on all the various leukocytes that bind to them is still fragmentary. While there is ample direct and indirect evidence that SLe^x is an important ligand in mediating the interaction of some cell types with the selectins, a thorough comparison of the distribution of SLe^x and other potential ligands on peripheral blood cells with their relative abilities to adhere to ELAM-1 and CD62 is still needed.

ROLES OF SELECTINS IN LEUKOCYTE RECRUITMENT *IN VIVO*

Although many details concerning the expression and cell type specificities of the selectins have been documented *in vitro,* information concerning their precise roles in leukocyte recruitment *in vivo* is just beginning to emerge. Based on the known activators of the expression of ELAM-1 (inflammatory cytokines, endotoxin) and CD62 (thrombin, histamine, H_2O_2) (see Table 2-2), their expression has been postulated to represent inflammatory and hemostatic responses to tissue injury, respectively.[2,50] Mel-14/LAM-1 also expressed participates in the recruitment of cells to sites of inflammation.[12,39] Indeed, it is well-documented that multiple adhesion proteins and their ligands are required for the process of leukocyte adhesion to and extravasation across endothelial cells.[1] The potential for chemotactic factors to attract specific classes of leukocytes add yet another element of regulation. Thus, it is not yet possible to predict precisely which adhesion receptors will govern the selective recruitment of specific leukocytes to sites of inflammation and tissue injury *in vivo*. Nonetheless, progress in understanding the roles of the selectins is being made on several fronts.

Mediation of Early Interactions of Leukocytes With Endothelial Cells

All three selectins appear to be involved in recruitment of neutrophils and other leukocytes to sites of inflammation. For the purpose of discussion, the recruitment of neutrophils to sites of inflammation can conveniently be divided into three steps (Figure 2-3): 1) initial attachment and "rolling" of cells on activated endothelium of postcapillary venules;[100,101] 2) activation of neutrophils and firm adhesion to the endothelium; and 3) extravasation of the cells into the surrounding tissue.[102,103]

Recent evidence from several laboratories suggests that the selectins participate in the initial adhesion or "rolling" of neutrophils on the activated endothelium.[15,16,104,105] By contrast, the subsequent arrest (firm adhesion) and extravasation of neutrophils is believed to be mediated by integrin receptors.[102,103,106,107] Indeed, antibodies to the common β-subunit (CD18) of the integrins Mac-1 and LFA-1 prevent firm adhesion and extravasation of neutrophils *in vivo*, but have no effect on the rolling of these cells on the

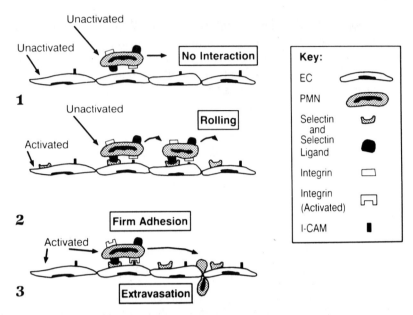

Figure 2-3. Generalized mechanism of recruitment of leukocytes to sites of inflammation. The model depicts the sequential action of selectin and integrin receptors in recruitment of leukocytes. The symbols for selectin and selectin ligand are interchangeable to reflect the evidence that CD62 and ELAM-1 on endothelial cells, and Mel-14/LAM-1 on leukocytes, each participate in leukocyte rolling under conditions in which they are expressed.

activated endothelium of postcapillary venules.[100] Moreover, neutrophils of patients with a deficiency in the CD18 integrins are incapable of firm adhesion to endothelial cells and resulting accumulation at sites of inflammation.[102]

In an elegant series of *in vitro* experiments using a flow cell system, Lawrence and Springer[104] provided evidence that CD62 could mediate rolling of neutrophils under conditions that mimic *in vivo* shear force of blood flow. CD62 and intercellular adhesion molecule-1 (ICAM-1), a counter ligand of the CD18 integrins LFA-1 and Mac-1, were embedded in an artificial phospholipid bilayer alone or in combination and examined for their ability to interact with their adhesion partners on human neutrophils. CD62 alone or CD62 with ICAM-1 produced rolling of neutrophils on the bilayer within the range of shear forces measured in postcapillary venules *in vivo,* while ICAM-1 alone caused neither rolling nor firm adhesion of neutrophils even at very low shear force. In contrast, when neutrophils were allowed to interact with the phospholipid bilayers under static conditions and then shear force was initiated, neutrophils developed such strong adhesion to ICAM-1 that they could not be displaced, but adhesive force generated with CD62 resulted in the same rolling of the cells observed under immediate shear force conditions. Finally, neutrophils rolling on bilayers containing both CD62 and ICAM-1 (but neither adhesion molecule alone) could be made to arrest and adhere firmly by including an activator, fMLP, in the bathing medium.

These experiments recapitulate the early steps in the recruitment of leukocytes to activated endothelial cells (see Figure 2-3) and provide a working hypothesis for the role of CD62 in leukocyte recruitment *in vivo,* specifically that CD62 can mediate the initial rolling of neutrophils, bringing them into close contact with the activated endothelial cells. They further suggest that the rolling of neutrophils on CD62 does not induce the activation and upregulation of the integrin receptors (LFA-1 and Mac-1) required for firm adhesion, as firm adhesion was not observed on a bilayer with CD62 and ICAM-1 under shear force without addition of the exogenous activator fMLP.[104]

Lorant et al.[51] have suggested that platelet activating factor (PAF) expressed on endothelial cells could initiate the activation of neutrophils in the context of the "rolling" of cells on activated endothelium mediated by CD62. CD62 and PAF are both expressed on endothelial cells following stimulation with thrombin, histamine, or oxidizing agents.[49,51,52] Moreover, adhesion of neutrophils to the activated endothelial cells could be inhibited by blocking either CD62 or the PAF receptor,[49,51] suggesting that the two receptors acted coordinately in mediating adhesion. Because the interaction of PAF with PAF receptor mediates adhesion by upregulation of the integrins LFA-1 and Mac-1,[108] Lorant et al.[51] propose that the initial tethering of neutrophils to endothelial cells by CD62 is required for the interaction of PAF with PAF receptor, resulting in the activation of the neutrophil integrins that mediate the firm adhesion of the cells to the blood vessel wall.

As mentioned previously, Mel-14/LAM-1 has been implicated in the recruitment of neutrophils to sites of inflammation *in vivo* because both antibody to Mel-14 and a Mel-14-IgG chimera can inhibit neutrophil emigration in inflammation models.[12,39,40] Using the same reagents, Ley et al.[16] and von Andrian et al.[15] have demonstrated inhibition of neutrophil rolling on postcapillary venules of intestinal mesentery induced by exteriorization of the tissue in preparation for intravital microscopy. In this preparation, the activation of the endothelium occurs spontaneously within minutes of exteriorization, and firm adhesion of cells is suppressed by administration of anti-CD18 antibodies. Although the receptor induced on endothelial cells that initiates neutrophil "rolling" has not been identified, the rapid onset of activation, involvement of oxygen-derived free radicals in the induction,[52,100,109] and the inhibition of leukocyte rolling by intravenous injection of sialidase[110] are all consistent with the expression and activity of CD62. Thus, it is possible that neutrophil rolling in this *in vivo* model may involve selectins on both the endothelial cell (CD62) and leukocyte Mel-14/LAM-1 side (see Figure 2-3).

Mel-14/LAM-1 has also been implicated in the CD18 independent adhesion to cytokine (IL-1β, TNFα)-stimulated endothelial cells,[13,14,105] as previously shown for ELAM-1.[17,41] Kishimoto et al.[13] have recently provided evidence that Mel-14/LAM-1 and ELAM-1 mediate the same CD18-independent adhesion step on binding of neutrophils to activated HUVECs *in vitro*. Under shear force in a flow cell system, rolling of neutrophils on cytokine-activated endothelial cells is observed.[105] The rolling is CD18-integrin-independent and is inhibited to the same extent by blocking antibodies to either Mel-14/LAM-1 or ELAM-1[105] (T. K. Kishimoto and C. W. Smith, personal communication). The results suggest that both Mel-14/LAM-1 and ELAM-1 participate in the CD18-independent rolling of neutrophils on cytokine-activated endothelial cells and that the two selectins may even serve as counter ligands for each other. These observations raise interesting possibilities for the subsequent steps of neutrophil activation, firm adhesion, and extravasation. In this regard, Lo et al.[111] have demonstrated that binding of neutrophils to ELAM-1 is sufficient to activate surface CD18 integrins. Moreover, upon activation of neutrophils, Mel-14/LAM-1 exhibits a transient increase in the affinity for its carbohydrate ligand[112] and is then rapidly shed.[113] The significance of these observations to the overall mechanism of neutrophil recruitment *in vivo* remains to be elucidated.

ELAM-1 Expression Associated With Acute and Chronic Inflammatory Disease

Although little is known about the role of ELAM-1 in mechanisms of leukocyte recruitment *in vivo*, its role in leukocyte-mediated disease is the best documented for any of the three selectins. Because ELAM-1 expression is induced *in vitro* by TNFα, IL-1β, and bacterial lipopolysaccharide (LPS) (see Chapter 4), this selectin has been implicated in the recruitment of neutrophils to sites of inflammation *in vivo* where these activators are likely present. Several recent reports have documented this prediction *in vivo*.

Using rat models of lung and skin injury induced by deposition of IgG immune complex, Mulligan et al.[44] demonstrated that anti-ELAM-1 antibodies could effect both a reduction in neutrophil accumulation and vascular damage. Anti-ELAM-1 also showed a similar reduction in neutrophil emigration in glycogen-induced peritoneal exudates. In the lung model, TNFα was implicated as the inflammatory cytokine that induces ELAM-1 expression on endothelial cells, as anti-TNFα antibodies also completely abolished neutrophil accumulation and lung injury.[44,114]

Because bacterial LPS can induce symptoms of septic shock and directly induces ELAM-1 on endothelial cells,[41] this selectin has been implicated in the recruitment of neutrophils in gram-negative sepsis resulting in multiple organ failure. Munro et al.[47] and Redl et al.[115] have demonstrated that LPS administration to baboons *in vivo* upregulates expression of ELAM-1 on postcapillary venules of lung, liver, kidney, and spleen, the organs often affected in this disease. Because administration of LPS to animals also causes a spike of both TNFα and IL1-β, these cytokines may be involved in upregulation of ELAM-1 on endothelial cells in addition to the direct induction by LPS.[47,115,116] Although the association of ELAM-1 expression in septic shock is clear, verification of its role in the pathologic recruitment of neutrophils to affected tissues awaits additional studies using anti-ELAM-1 or other inhibitors to block neutrophil accumulation.

ELAM-1 has also been found to be expressed in chronic inflammatory disease including rheumatoid arthritis[117,118] and various chronic skin inflammations such as psoriasis and allergic cutaneous inflammation.[45,46,98,99] The observed expression of ELAM-1 throughout the course of chronic inflammation is in marked contrast to the transient expression (2 to 8 hours) seen during cytokine-induced expression *in vitro*,[41] or in LPS and immune complex-induced expression *in vivo*.[44,47] As yet, the basis for the sustained expression of ELAM-1 has not been elucidated.

In chronic inflammatory disease, ELAM-1 expression is not solely associated with recruitment of neutrophils. Indeed, in psoriasis and other skin inflammations,[45,99] ELAM-1 expression is associated with the accumulation of a subset of T lymphocytes (HECA-452 positive, memory phenotype-enriched). In a model of allergic contact dermatitis (poison ivy extract), accumulation of leukocytes was initiated in 24 hours, and the leukocytes recruited included lymphocytes, monocytes, and macrophages, but not neutrophils. In contrast, using different allergens (dust mite and pollen), Kyan-Aung et al.[98] and Leung et al.[119] found predominant accumulation of eosinophils and/or neutrophils in two to six hours after administration of allergen. The basis for the differential recruitment of T cells and monocytes in response to one allergen and the recruitment of eosinophils and neutrophils in response to another is not clear at present. Because ELAM-1 binds each of these cell types *in vitro,* it is a likely candidate for participating in their recruitment to sites of skin inflammation. However, these reports clearly point out that the specificity of this selectin for its carbohydrate ligand is not necessarily the dominant factor in the cell type selectivity in skin inflammation.

SUMMARY

Although there has been rapid progress in understanding the structures of the selectins and their carbohydrate ligands, there are still major gaps in understanding the detailed ligand specificities of all three selectins, and to what degree ligand specificity determines the cell type selectivities of the selectins in recruitment of leukocytes to sites of inflammation and tissue injury. There is an emerging consensus that all three selectins are involved in the early adhesion events (rolling) of leukocytes on activated endothelial cells of postcapillary venules. However, this evidence is still fragmentary, and additional information is needed to understand how multiple adhesion receptors and cytokines cooperate to recruit leukocytes in response to different inflammatory signals. The challenge presented to solve these questions ensures that the field opened by the identification of the selectins as a homologous gene family will remain a fertile and productive one in the coming years.

REFERENCES

1. Springer, T.A. (1990) *Annu. Rev. Cell Biol.* 6, 359–402.
2. Geng, J.G., Bevilacqua, M.P., Moore, K.L., McIntyre, T.M., Prescott, S.M., Kim, J.M., Bliss, G.A., Zimmerman, G.A., and McEver, R.P. (1990) *Nature* 343, 757–760.
3. Stoolman, L.M. (1989) *Cell* 56, 907–910.
4. Brandley, B.K., Swiedler, S.J., and Robbins, P.W. (1990) *Cell* 63, 861–863.
5. Springer, T.A., and Lasky, L.A. (1991) *Nature* 349, 196–197.
6. Feizi, T. (1991) *Treds Biochem Sci* 16, 84–86.
7. Siegelman, M.H. (1991) *Current Biol.* 1, 125–128.
8. Pober, J.S. and Cotran, R.S. (1991) *Lab. Invest.* 64, 301–305.
9. Yednock, T.A. and Rosen, S.D. (1989) *Adv. Immunol.* 44, 313–378.
10. Kansas, G.S., Spertini, O., and Tedder, T.F. (1991) *Cellular and Molecular Mechanisms of Inflammation: Vascular Adhesion Molecules* (Cochrane, C., and Gimbrone, M., eds.) pp.1-40., Academic Press, NY
11. Griffin, J.D., Spertini, O., Ernst, T.J., Belvin, M.P., Levine, H.B., Kanakura, Y., and Tedder, T.F. (1990) *J. Immunol.* 145, 576–584.
12. Watson, S.R., Fennie, C., and Lasky, L.A. (1991) *Nature* 349, 164–167.
13. Kishimoto, T.K., Warnock, R.A., Jutila, M.A., Butcher, E.C., Lane, C., Anderson, D.C., and Smith, C.W. (1991) *Blood* 78, 805–811.
14. Hallmann, R., Jutila, M.A., Smith, C.W., Anderson, D.C., Kishimoto, T.K., and Butcher, E.C. (1991) *Biochem Biophys Res Commun* 174, 236–243.
15. von Andrian, U.H., Chambers, J.D., McEvoy, L.M., Bargatze, R.F., Arfors, K.E., and Butcher, E.C. (1991) *Proc. Nat. Acad. Sci. USA* 88, 7538–7542.
16. Ley, K., Gaehtgens, P., Fennie, C., Singer, M.S., Lasky, L.A., and Rosen, S.D. (1991) *Blood* 77, 2553–2555.
17. Bevilacqua, M.P., Stengelin, S., Gimbrone, M.A., and Seed, B. (1989) *Science* 243, 1160–1161.
18. Polte, T., Newman, W., and Gopal, T.V. (1989) *Nucleic Acids Res.* 18, 1083.

Acknowledgments: I am indebted to all the friends and colleagues who were kind enough to send preprints, submit manuscripts or communicate unpublished observations to make this a timely and up-to-date review: Dr. Karl Arfors, Dr. Brian Brandley, Dr. Sen-Itiroh Hakomori, Dr. John Harlan, Dr. Reggi Kanagi, Dr. Takashi K. Kishimoto, Dr. Roy Lobb, Dr. John Lowe, Dr. Brian Nickoloff, Dr. Jordan Pober, Dr. Wayne Smith, Dr. Timothy Springer, Dr. Stuart Swiedler, Dr. Thomas Tedder, Dr. Uli von Andrian, Dr. Peter Ward, and Dr. Guy Zimmerman.

19. Johnston, G.I., Kurosky, A., and McEver, R.P. (1989) *J. Biol. Chem.* 264, 1816–1823.
20. Tedder, T.F., Isaacs, C.M., Ernst, T.J., Demetri, G.D., Adler, D.A., and Disteche, C.M. (1989) *J. Exp. Med.* 170, 123–133.
21. Bowen, B.R., Nguyen, T., and Lasky, L.A. (1989) *J. Cell Biol.* 109, 421–427.
22. Camerini, D., James, S.P., Stamenkovic, J., and Seed, B. (1989) *Nature* 342, 78–82.
23. Lasky, L.A., Singer, M.S., Yednock, T.A., Dowbenko, D., Fennie, C., Rodriguez, H., Nguyen, T., Stachel, S., and Rosen, S.D. (1989) *Cell* 56, 1045–1055.
24. Siegelman, M.H., van de Rijn, M., and Weissman, I.L. (1989) *Science* 243, 1165–1172.
25. Hession, C., Osborn, L., Goff, D., Chi-Rosso, G., Vassallo, C., Pasek, M., Pittack, C., Tizard, R., Goelz, S., McCarthy, K., Hopple, S., and Lobb, R. (1990) *Proc. Natl. Acad. Sci. USA* 87, 1673–1677.
26. Siegelman, M.H., Cheng, I.C., and Weissman, I.L., Wakeland, E.K. (1990) *Cell* 61, 611–622.
27. Tedder, T.F., Penta, A.C., Levine, H.B., and Freedman, A.S. (1990) *J. Immunol.* 144, 532–540.
28. Drickamer, K. and McCreary, V. (1987) *J. Biol. Chem.* 262, 2582–2589.
29. Drickamer, K. (1988) *J. Biol. Chem.* 263, 9557–9560.
30. Johnston, G.I., Cook, R.G., and McEver, R.P. (1989) *Cell* 56, 1033–1044.
31. Stoolman, L.M. and Rosen, S.D. (1983) *J. Cell Biol.* 96, 722–729.
32. Rosen, S.D., Singer, M.S., Yednock, T.A., and Stoolman, L.M. (1985) *Science* 228, 1005–1007.
33. Stoolman, L.M., Yednock, T.A., and Rosen, S.D. (1987) *Blood* 70, 1842–1850.
34. Rosen, S.D., Chi, S.I., True, D.D., Singer, M.S., and Yednock, T.A. (1989) *J. Immunol.* 142, 1895–1902.
35. Imai, Y., Singer, M.S., Fennie, C., Lasky, L.A., and Rosen, S.D. (1991) *J. Cell Biol.* 113, 1213–1221.
36. Gamble, J.R., Skinner, M.P., Berndt, M.C., and Vadas, M.A. (1990) *Science* 249, 414–417.
37. Skubitz, K.M., and Snook, R.W. (1987) *J. Immunol.* 139, 1631–1639.
38. Asada, M., Furukawa, K., Kantor, C., Gahmberg, C.G., and Kobata, A. (1991) *Biochemistry* 30, 1561–1571.
39. Lewinsohn, D.M., Bargatze, R.F., and Butcher, E.C. (1987) *J. Immunol.* 138, 4313–4321.
40. Jutila, M.A., Rott, L., Berg, E.L., and Butcher, E.C. (1989) *J. Immunol.* 143, 3318–3324.
41. Bevilacqua, M.P., Pober, J.S., Mendrick, D.L., Cotran, R.S., and Gimbrone, M.A. (1987) *Proc. Natl. Acad. Sci. USA* 84, 9238–9242.
42. Leeuwenberg, J.F.M., VonAsmuth, E.J.U., Jeunhomme, T.M.A.A., and Buurman, W.A. (1990) *J. Immunol.* 145, 2110–2114.
43. Matis, W.L., Lavker, R.M., and Murphy, G.F. (1990) *J. Invest. Dermatol.* 94, 492–495.
44. Mulligan, M.S., Varani, J., Dame, M.K., Lane, C.L., Smith, C.W., Anderson, D.C., and Ward, P.A. (1991) *J. Clin. Invest.* 88, 1396–1406.

45. Picker, L.J., Kishimoto, T.K., Smith, C.W., Warnock, R.A., and Butcher, E.C. (1991) *Nature* 349, 796–799.
46. Griffiths, C.E., Barker, J.N., Kunkel, S., and Nickoloff, B.J. (1991) *Brit. J. Dermatol.* 124, 519–526.
47. Munro, J.M., Pober, J.S., and Cotran, R.S. (1991) *Lab. Invest.* 64, 295–299.
48. Larsen, E., Palabrica, T., Sajer, S., Gilbert, G.E., Wagner, D.D., Furie, B.C., and Furie, B. (1990) *Cell* 63, 467–474.
49. Watanabe, M., Yagi, M., Omata, M., Hirasawa, N., Mue, S., Tsurufuji, S., and Ohuchi, K. (1990) *Br. J. Pharmacol.* 102, 239–245.
50. Toothill, V.J., Van Mourik, J.A., Niewenhuis, H.K., Metzelaar, M.J., and Pearson, J.D. (1990) *J. Immunol.* 145, 283–291.
51. Lorant, D.E., Patel, K.D., McIntyre, T.M., McEver, R.P., Prescott, S.M., and Zimmerman, G.A. (1991) *J. Cell Biol.* In press.
52. Patel, K.D., Zimmerman, G.A., Prescott, S.M., McEver, R.P., and McIntyre, T. (1991) *J. Cell Biol.* 112, 749–759.
53. Hattori, R., Hamilton, K.K., Fugate, R.D., McEver, R.P., and Sims, P.J. (1989) *J. Biol. Chem.* 264, 7768–7771.
54. Rosen, S.D., Singer, M.S., and Yednock, T.A. (1985) *Science* 288, 1005–1007.
55. Polley, M.J., Phillips, M.L., Wayner, E., Nudelman, E., Singhal, A.K., Hakomori, S.I., and Paulson, J.C. (1991) *Proc. Natl. Acad. Sci. USA* 88, 6224–6228.
56. Phillips, M.L., Nudelman, E., Gaeta, F.C.A., Perez, M., Singhal, A.K., Hakomori, S.I., Paulson, J.C. (1990) *Science* 250, 1130-1132.
57. Lowe, J.B., Stoolman, L.M., Nair, R.P., Larsen, R.D., Berhend, T.L., and Marks, R.M. (1990) *Cell* 63, 475–484.
58. Goelz, S.E., Hession, C., Goff, D., Griffiths, B., Tizard, R., Newman, B., Chi-Rosso, G., and Lobb, R. (1990) *Cell* 63, 1349–1356.
59. Walz, G., Aruffo, A., Kolanus, W., Bevilacqua, M.P., and Seed, B. (1990) *Science* 250, 1132–1135.
60. Tiemeyer, M., Swiedler, S.J., Ishihara, M., Moreland, M., Schweinbruber, H., Hirtzer, P., and Brandley, B.K. (1991) *Proc. Natl. Acad. Sci. USA* 88, 1138–1142.
61. Mizoguchi, A., Takasaki, S., Maeda, S., and Kobata, A. (1984) *J. Biol. Chem.* 259, 11949–11957.
62. Fukuda, M., Spooncer, E., Oates., J.E., Dell, A., and Klock, J.C. (1984) *J. Biol. Chem.* 259, 10925–10935.
63. Tetteroo, P.A.T., van't Veer, M.B., Tramp, J.F., and Von dem Borne, A.E. (1984) *Int. J. Cancer* 33, 355–358.
64. Fukuda, M., Bothner, B., Ramsamooj, P., Dell, A., Tiller, P.R., Varki, A., and Klock, J.C. (1985) *J. Biol. Chem.* 260, 12957–12967.
65. Fukuda, M.N., Dell, A., Oates, J.E., Wu, P., Klock, J., and Fukuda, M. (1985) *J. Biol. Chem.* 260, 1067–1079.
66. Spitalnik, S.L., Spitalnik, P.F., Civin, C.I., Ball, E., Schwartz, J.F., and Ginsburg, V. (1986) *Exp. Hematol.* 14, 643–647.
67. Majdic, O.M., Liszka, K., Lutz, D., and Knapp, W. (1981) *Blood* 58, 1127–1133.
68. Gooi, H.C., Thorpe, S.J., Hounsell, E.F., Rumpold, H., Kraft, D., Foster, O., and Feizi, T. (1983) *Eur. J. Immunol.* 13, 306–312.
69. Skubitz, K.M., Pessano, S., Bottero, L., Ferraro, D., Rovera, G., and August, J.T. (1983) *J. Immunol.* 131, 1882–1888.

70. Stockinger, H., Majdic, O.M., Liszka, K., Aberer, W., Bettelheim, P., Lutz, D., and Knapp, W. (1984) *J. Natl. Cancer Inst.* 73, 7–11.
71. Campbell, C. and Stanley, P. (1984) *J. Biol. Chem.* 259, 11208–11214.
72. Stanley, P., and Atkinson, P.H. (1988) *J. Biol. Chem.* 263, 11374–11381.
73. Potvin, B., Kumar, R., Howard, D.R., and Stanley, P. (1990) *J. Biol. Chem.* 265, 1615–1622.
74. Lowe, J.B., Kukowska-Latallo, J., Nair, R.P., Larsen, R.D., Marks, R.M., Macher, B.A., Kelly, R.J., and Ernst, L.K. (1991) *J. Biol. Chem.* In press.
75. Corral, L., Singer, M.S., Macher, B.A., and Rosen, S.D. (1990) *Biochem. Biophys. Res. Commun.* 172, 1349–1356.
76. Moore, K.L., Varki, A., and McEver, R.P. (1991) *J. Cell Biol.* 112, 491–499.
77. Hakomori, S.I. and Kobata, A. (1974) in *The Antigens* (Sela, M., ed.) pp. 79–140, Academic Press, NY.
78. Solter, D. and Knowles, B.B. (1978) *Proc. Natl. Acad. Sci. USA* 75, 5565–5569.
79. Gooi, H.C., Feizi, T., Kapadaia, A., Knowles, B.B., Solter, D., and Evans, M.J. (1981) *Nature* 292, 156–158.
80. Ohmori, K., Toneda, T., Ishihara, G., Shigeta, K., Hirashima, K., Kanai, M., Itai, S., Sasaoki, T., Arii, S., Arita, H., and Kannagi, R. (1989) *Blood* 74, 255–261.
81. Tabilio, A., DelCanzio, M.C., Henri, A., Guichard, J., Mannoni, P., Civin, C.I., Testa, U., Rochant, H., Vainchenker, W., and Brenton-Gorius, J. (1984) *Br. J. Haematol.* 58, 697–710.
82. Terasaki, P.I., Hirota, M., Fukushima, K., Wakisaka, A., and Iguro, T. (1988) U.S. Patent #4,752,569.
83. Holmes, E.H., Ostrander, G.K., and Hakomori, S.I. (1986) *J. Biol. Chem.* 261, 3737–3743.
84. Garin-Chesa, P., and Rettig, W.J. (1989) *Am. J. Path.* 134, 1315–1327.
85. Sakamoto, J., Watanabe, T., Tolumaru, T., Takagi, H., Nakazato, H., and Lloyd, K.O. (1989) *Cancer Res.* 49, 745–752.
86. Lee, N., Wang, W.C., and Fukuda, M. (1990) *J. Biol. Chem.* 265, 20476–20487.
87. Macher, B.A., and Beckstead, J.H. (1990) *Leukemia Res.* 14, 119–130.
88. Mitsakos, A., Hanisch, F.G., and Uhlenbruck, G. (1988) *Biol. Chem. Hoppe Seyler* 369, 661–665.
89. Mollicone, R., Gibaud, A., Francois, A., Ratcliffe, M., and Oriol, R. (1990) *Eur. J. Biochem.* 191, 169–176.
90. Greenwell, P., Ball, M.G., and Watkins, W.M. (1983) *FEBS Lett.* 164, 314–317.
91. Prieels, J.P., Monnom, D., Dolmans, M., Beyer, T.A., and Hill, R.L. (1981) *J. Biol. Chem.* 256, 10456–10463.
92. Stroup, G.B., Anumula, K.R., Kline, T.F., and Caltabiano, M.M. (1990) *Cancer Res.* 50, 6787–6792.
93. Johnson, P.H. and Watkins, W.M. (1982) *Biochem. Soc. Trans.* 10, 445–446.
94. Johnson, P.H., Yates, A.D., and Watkins, W.M. (1981) *Biochem. Biophys. Res. Comm.* 100, 1611–1618.
95. Kukowska-Latallo, J.F., Larsen, R.D., Rajan, V.P., and Lowe, J.B. (1990) *Genes Dev.* 4, 1288–1303.

96. Thorpe, S.J. and Feizi, T. (1984) *Biosci. Rep.* 4, 673–685.
97. Lobb, R.R., Chi-Rosso, G., Leone, D.R., Rosa, M.D., Bixler, S., Newman, B.M., Luhowskyj, S., Benjamin, C.D., Dougas, I.G., Goelz, S.E., Hession, C., and Chow, E.P. (1991) *J. Immunol.* 147, 124–129.
98. Kyan-Aung, U., Haskard, D.O., Poston, R.N., Thornhill, M.H., and Lee, T.H. (1991) *J. Immunol.* 146, 521–528.
99. Shimizu, Y., Shaw, S., Graber, N., Gopal, T.V., Horgan, K.J., Van Seventer, G.A., and Newman, W. (1991) *Nature* 349, 799–802.
100. Arfors, D.E., Lundberg, C., Lindborm, L., Lundberg, K., Beatty, P.G., and Harlan, J.M. (1987) *Blood* 69, 338–340.
101. Smith, C.W., Rothlein, R., Hughes, B.J., Mariscalco, M.M., Schmalsteig, F.C., and Anderson, D.C. (1988) *J. Clin. Invest.* 82, 1746–1756.
102. Anderson, D.C. and Springer, T.A. (1987) *Ann. Rev. Med.* 38, 175–194.
103. Larson, R.S. and Springer, T.A. (1990) *Immunol. Rev.* 114, 181–217.
104. Lawrence, M.B., and Springer, T.A. (1991) *Cell* 65, 859–873.
105. Smith, C.W., Kishimoto, T.K., Abbass, O., Hughes, B., Rothlein, R., McIntire, L.V., Butcher, E., and Anderson, D.C. (1991) *J. Clin. Invest.* 87, 609–618.
106. Smith, C.W., Marlin, S.D., Rothlein, R., Toman, C., and Anderson, D.C. (1989) *J. Clin. Invest.* 83, 2008–2017.
107. Luscinskas, F.W., Cybulsky, M.I., Kiely, J.M., Peckins, C.S., Davis, V.M., and Gimbrone, M.A. (1991) *J. Immunol.* 146, 1617–1625.
108. Zimmerman, G.A., McIntyre, R.M., Mehra, M., and Prescott, S.M. (1990) *J. Cell Biol.* 110, 529–540.
109. Del Maestro, R.F., Planker, M., and Arfors, K.E. (1982) *Int. J. Microcirc. Clin. Exp.* 1, 105–120.
110. Atherton, A., and Born, G.V.R. (1973) *J. Physiol.* 234, 66P–67P.
111. Lo, S.K., Lee, S., Ramos, R.A., Lobb, R., Rosa, M., Chi-Rosso, G. and Wright, S.D. (1991) *J. Exp. Med.* 173, 1493–1500.
112. Spertini, O., Kansas, G.S., Munro, J.M., Griffin, J.D., and Tedder, T.F. (1991) *Nature* 349, 691–694.
113. Kishimoto, T.K., Jutila, M.A., Berg, E.L., and Butcher, E.C. (1989) *Science* 245, 1238–1241.
114. Warren, J.S., Yabroff, K.R., Remick, D.G., and Kunkel, S.L. (1989) *J. Clin. Invest.* 84, 1873–1882.
115. Redl, H., Dinges, H.P., Buurman, W.A., van der Linden, C.J., Pober, J.S., Cotran, R.S., and Schlag, G. (1991) *Am. J. Path.* 139, 461–466.
116. Hinshaw, L.B., Tekamp-Olson, P., Chang, A.C., Lee, P.A., Taylor, F.B., Murray, C.R., Peer, G.T., Emerson, T.E., Passey, R.B., and Kuo, G.C. (1990) *Circ. Shock* 30, 279–292.
117. Corkill, M.M., Kirkham, B.W., Haskard, D.O., Barbatis, C., Gibson, T., and Panayi, G.S. (1991) *J. Exp. Med.* In press.
118. Koch, A.E., Burrows, J.C., Haines, K.G., Carlos, T.M., Harlan, J.M., and Leibovich, S.J. (1991) *Lab. Invest.* 64, 313–320.
119. Leung, D.Y.M., Pober, J.S., and Cotran, R.S. (1991) *J. Clin. Invest.* 87, 1805–1809.
120. Hamburger, S.A., and McEver, R.P. (1990) *Blood* 75, 550–554.
121. Rice, G.E., and Bevilacqua, M.P. (1989) *Science* 246, 1303–1306.

CHAPTER 3

The Homing Receptor (LECAM 1/L Selectin): A Carbohydrate-Binding Mediator of Adhesion in the Immune System

Laurence A. Lasky

Leukocytic or white blood cells, given their role as protectors of the organism, must be highly mobile so that they are able to migrate quickly to regions of insult in order to deal with invading pathogens. These cells must then undergo a hasty metamorphosis that results in a change from a mobile state to a highly adhesive one. These adhesive cells then adhere to the endothelium adjacent to regions under attack, after which they become activated advocates of the immune system, resulting in migration toward and ultimate destruction of the pathogen. Thus, these cells must maintain, through a complex combination of gene expression and protein regulation, either nonadhesive or adhesive states at appropriate times in order to insure the highest degree of immunologic surveillance to the organism.

The importance of this so-called inflammatory response is underlined by the very limited life expectancy of individuals that have one or more defects in this protective pathway. For example, individuals with the leukocyte adhesion deficiency (LAD) syndrome have a simple defect in the synthesis of one of the molecules involved in the adhesive interaction between leukocytes and the endothelium, the beta 2 integrin chain.[1-4] The absence of this single protein component causes a lack of cell surface heterodimeric integrin expression and results in the inability of neutrophils to properly adhere to the vessel wall and migrate to sites of, for example, bacterial infection. This often leads to a series of virulent bacterial infections that can cause the early death of the individual. These findings underline the importance of the inflammatory response to protection of the organism and are consistent with the critical role of leukocyte adhesion in this protective response. In addition, they have stimulated a number of animal studies[4a] that indicate a clinically relevant role for antibodies directed against leukocyte integrin subunits.

In the 1980s, an explosion in knowledge regarding the cell and molecular biology of leukocyte adhesion occurred. Perhaps the most interesting aspect of these discoveries was a description of three of the families of molecules involved with leukocyte adhesion during the inflammatory response.[4] The first of these families, mentioned above, is encompassed by the heterodimeric alpha and beta integrin chains. The second of these families corresponds to some of the glycoprotein ligands for the integrins and has been referred to as the *immunoglobulin superfamily of adhesion molecules* (see Chapter 1 for a review of these two families). The final, and most recently discovered, family is unique among cell adhesion molecules because it accomplishes adhesion by means of protein–carbohydrate, rather than protein–protein, interactions. This family, the LECCAM (lectin cell adhesion molecule) or selectin family, contains three members: the homing receptor (LECAM 1, L selectin), ELAM-1 (LECAM 2, E selectin) and GMP-140/PADGEM (LECAM 3, P selectin).[5,6] While molecular descriptions for all three of these glycoproteins were communicated virtually simultaneously, the paradigm for this unique type of adhesion molecule, and the subject of this chapter, is the homing receptor.

HISTORICAL PERSPECTIVE

One of the most important trafficking pathways of the immune system is that traversed by B and T lymphocytes as they migrate from the bloodstream to the peripheral lymphoid tissue.[7-13] This is a critical migratory conduit, as it allows for cells with a diverse spectrum of antigenic specificities to encounter, with the highest probability, external, and often harmful, antigenic stimuli. Early *in vivo* experiments by Gowans, Ford and others suggested that lymphocytes isolated from peripheral lymphoid tissues tended to migrate back to this type of lymphoid organ, while those cells taken from other lymphoid regions tended to migrate back to these realms, implying some type of organ-specific migration or "homing" capacity.[4,9] It was theorized early on that the induction of this organ- or region-specific homing might have been due to adhesion molecules on the lymphocyte cell surface that were specialized to adhere to ligands expressed only in one type of lymphoid organ or another.[13] However, the relatively primitive state of cell and molecular biology and immunology at that time prevented an accurate molecular description of such adhesion molecules, and the observation of organ-specific homing remained an interesting, although poorly explained, phenomenon.

Two major breakthroughs allowed for the description of at least one of the molecules involved in organ-specific lymphocyte homing. The first consisted of an *in vitro* assay that replicated the adhesive interactions between lymphocytes and the endothelium of lymphoid organs. In this assay, named the Stamper-Woodruff assay after its developers, frozen sections of various lymphoid organs were incubated with lymphocytes, the sections were washed, and the degree of specific binding between the added lymphocytes and the specialized high walled endothelium of the postcapillary venules of these organs was determined.[14,15] An interesting initial finding with this assay was that the adherence of lymphocytes to peripheral lymph node endothelial venules was entirely calcium-dependent (see below). In addition, the resultant adhesive interactions seemed to specifically represent those found *in vivo,* and, therefore, allowed for the development of immunologic reagents that could detect the adhesive molecules involved with these specific interactions. This led to the second breakthrough, where a rat monoclonal antibody, termed the *MEL-14 antibody,* directed against a 90-kilodalton (kD) glycoprotein on the murine lymphocyte cell surface, was found to specifically block the adherence of lymphocytes to the postcapillary venule endothelium of peripheral, but not Peyer's patch (gut), lymphoid organs.[16,17] This result demonstrated two important concepts. The first was that a specific glycoprotein on the lymphocyte cell surface was, at least in part, responsible for the adhesion of these cells to the lymphoid endothelium. The second was that this adhesion appeared to be specific for one type of lymphoid organ, the peripheral lymph node, suggesting that the ligand for this adhesion molecule may have been expressed in a lymphoid tissue-specific manner. The latter result was consistent with previous theories of potential mechanisms for organ-specific homing and led to the name "homing receptor" for the 90-kD adhesive glycoprotein recognized by the MEL-14 monoclonal antibody.

At this time, Rosen, Stoolman and colleagues were utilizing the same *in vitro* cell-binding assays to investigate the possibility that the adhesion between lymphocytes and peripheral lymphoid endothelium was due to protein–carbohydrate interactions. While considered somewhat heretical at the time, these investigators provided clear evidence in the early 1980s that some charged monomeric carbohydrates, such as mannose-6-phosphate, were able to specifically inhibit the adhesive interactions between lymphocytes and peripheral lymph node endothelium, although these carbohydrates were effective only at millimolar concentrations.[18] Such inhibition was not found with Peyer's patch endothelium, consistent with the possibility that a different adhesion system was utilized in this lymphoid organ. These investigators went on to show that polymers of mannose-6-phosphate (polyphosphomannan ester or PPME) or fucose-4-sulfate (fucoidin) from natural sources (yeast and algae, respectively) were effective inhibitors of lymphocyte-peripheral lymph node (lymphocyte-pln) endothelium binding at much lower concentrations (micromolar), further strengthening the concept that carbohydrates may have been involved in this adhesive interaction.[19,20] It was also shown that treatment of the lymphocyte, but not the endothelium, with such blocking carbohydrates inhibited adhesion, consistent with the expression of a carbohydrate-binding molecule on the lymphocyte, but not endothelial, cell surface. In addition, fluorescent beads coated with PPME were found to bind to the lymphocyte, further strengthening the possibility that the carbohydrate-binding receptor was on the lymphocyte surface. The calcium dependence of this PPME bead binding was noted as an interesting correlation with the calcium dependence of lymphocyte–endothelial binding in the Stamper-Woodruff assay.[21] Finally, treatment of peripheral lymph node, but not Peyer's patch, sections with the sialic acid-removing enzyme sialidase (neuraminidase) inhibited the endothelial binding of lymphocytes *in vitro* and *in vivo*,[21a] consistent with the possibility that this interaction was, at least in part, mediated by the carbohydrate sialic acid.

The results of the blocking of peripheral lymph node endothelium binding by lymphocytes with both the MEL-14 monoclonal antibody and carbohydrates suggested a possible relationship between the 90-kD homing receptor recognized by MEL-14 and a potential carbohydrate-binding adhesion molecule on the lymphocyte cell surface. The MEL-14 antibody and carbohydrate-blocking studies were united when it was shown that this antibody could block the binding of PPME-coated beads to the lymphocyte cell surface. It was also shown that cells selected for high levels of MEL-14 antigen expression also bound higher levels of PPME and adhered to pln endothelium more dramatically.[22] These results were consistent with the possibility that the 90-kD antigen recognized by the MEL-14 antibody bound to carbohydrates such as PPME and, because both the antibody and carbohydrates blocked lymphocyte–endothelial binding, that this glycoprotein adhered to the endothelial surface by means of protein–carbohydrate interactions. While this idea was extremely provocative, the field awaited the next major step toward demonstrating the role of carbohydrate recognition in lymphocyte–endothelium adhesion: the cDNA cloning of the homing receptor.

cDNA CLONING OF THE MURINE AND HUMAN HOMING RECEPTOR

N-terminal micro-sequencing of homing receptor antigen isolated by MEL-14 antibody affinity chromatography allowed two groups to simultaneously clone the cDNAs encoding the murine homing receptor.[23,24] Comparison of the encoded amino acid sequence with previously reported protein sequences revealed that the entire extracellular domain of the homing receptor was constructed from a number of motifs that were derived from other proteins. The N-terminus of the homing receptor encoded a signal sequence utilized for protein secretion into the endoplasmic reticulum. Immediately following this domain was a 116-amino acid motif that was homologous to the type C or calcium dependent carbohydrate binding proteins or lectins. While the overall homology in this region with these other lectins was relatively low (~25%–30%), there was a high degree of conservation of a subset of amino acids (the so-called "Drickamer motif" residues[25]) that are predominant in a diversity of calcium-dependent lectins. In addition, the disulfide bridges were also conserved with other lectins in this family. These homology comparisons, in addition to the previous hypotheses concerning the possible relationships between the homing receptor and a carbohydrate-binding receptor, were thus consistent with a physiologically important, carbohydrate-binding role for this domain. Following this motif was a 33-amino acid region that was homologous to the epidermal growth factor motif found in a wide variety of proteins. The degree of conservation here was also quite low, but potentially structural amino acids such as cysteines and glycines were highly conserved. Following this motif were two identical copies, at both the amino acid and nucleic acid level, of a 62-amino acid motif homologous to the short consensus repeat found in a wide array of complement-binding proteins.[26] This region was followed by a short region that led to a highly hydrophobic transmembrane anchor domain. Finally, a short, charged cytoplasmic domain was seen. In summary, the homing receptor was found to contain a number of protein motifs; potentially the most interesting and relevant is an N-terminal domain highly related to carbohydrate-binding (lectin) proteins.

The discovery of a C-type (calcium-dependent) lectin at the N-terminus of the homing receptor vindicated previous carbohydrate blocking data and, together with this data, provided the first clear evidence that cell adhesion in the immune system may be accomplished by protein–carbohydrate interactions. In addition to the carbohydrate blocking studies, these results were entirely consistent with the previously-noted calcium dependence of lymphocyte–lymph node endothelium and lymphocyte–PPME binding, as the carbohydrate binding by this type of lectin is entirely calcium-dependent. Indeed, at this time, the data were entirely consistent with this adhesion being solely due to the interaction of the homing receptor lectin domain with an endothelial-specific carbohydrate(s). As discussed below, however, the problem is somewhat more complicated than this. Suffice it to say, however, that what began as simple experiments demonstrating blocking of cell adhesion with monomeric sugars led to the discovery of a new type of adhesion molecule that appeared to regulate lymphocyte adherence through protein–carbohydrate interactions.

In contrast to the clear-cut results that were obtained with the MEL-14 antibody in the murine system, the identity of the human homing receptor was quite confusing. Butcher and colleagues reported in a series of papers that an antigen that they identified as the Hermes glycoprotein by antibody studies was the human equivalent of the murine homing receptor.[7] This identity was based on similar molecular weights (~90 kD), the ability to pre-clear murine homing receptor with a polyclonal antibody against the Hermes glycoprotein, and adhesion blocking studies using the previously described frozen section assay. This latter assay gave somewhat conflicting results depending upon the type of lymphoid tissue examined and the nature of the anti-Hermes antibody (monoclonal or polyclonal) used for the blocking studies. In addition, the tissue distribution of the Hermes glycoprotein was completely different from that found previously for the murine homing receptor, raising further doubts about its homology with the murine homing receptor. As with so many confusing issues in cell biology, the issue was clarified by the molecular cloning of the cDNAs encoding the human Hermes antigen and the human equivalent of the murine homing receptor.[27-30] These data indicated that the preliminary data concerning the homology of the Hermes glycoprotein and the murine homing receptor were incorrect. Thus, the Hermes antigen was found to encode a completely different glycoprotein that was homologous to cartilage link proteins and that was subsequently shown to be a hyaluronic acid receptor. This conclusion did not support the hypothesis that the Hermes antigen was a specific homing receptor, but instead was consistent with the possibility that this protein may function in a more nonspecific manner in cell adhesion. In contrast, the human homologue of the murine homing receptor was found to bear a high degree of sequence homology with its murine counterpart. The human receptor contained a lectin domain, an epidermal growth factor (egf) domain, two nonidentical copies of a complement binding-like motif, a transmembrane anchor, and a short cytoplasmic tail. The relative homologies were quite high in the lectin and egf motifs (~85%) and somewhat lower in the complement binding-like motifs (~70%-80%). Interestingly, the transmembrane domain showed almost complete homology between the human and mouse sequences (1 amino acid change in 39 residues [~97%] include regions before and after the ~20 residue transmembrane motif), suggesting that this region may fulfill more complex functions than mere plasma membrane anchoring. In addition, the cytoplasmic tail of both the human and murine homologues both appeared to contain at least one potential protein kinase C-dependent phosphorylation site, consistent with the possibility that this high degree of conservation in the transmembrane and surrounding regions may have been involved with interactions with kinases. The high degree of overall sequence homology between the human and murine homing receptors was consistent with previous trafficking and cell-binding assays that demonstrated a prominent level of conservation between the lymphocyte adhesive and migratory mechanisms of human and mouse. In conclusion, it appeared that the adhesive pathway mediated by the homing receptor was a relatively early physiologic solution that remained little changed during the evolution from mouse to man.

An additional important finding derived from the cloning of the human homing receptor was the demonstration that it was homologous to a previously described lymphocyte surface antigen of unknown identity that was recognized by the Leu 8 and TQ 1 monoclonal antibodies.[28,31] These results were interesting because they allowed for a reappraisal of the previous work done with these antibodies. This work suggested that different T-cell subsets appeared to express different levels of the homing receptor.[32,33] For example, the naive subset appeared to show uniformly high expression of the homing receptor while the memory subset appeared to show a bi-modal distribution of expression.[31,33] These results were consistent with the possibility that the homing receptor mediated trafficking of different T-cell subsets, depending upon their phenotype. More recent work with the Leu 8 antibody has shown that skin lymphocytes appeared to also express high levels of this adhesion molecule, suggesting a possible role for this glycoprotein in skin trafficking. Additional antibodies against the human homing receptor have been produced (i.e., the LAM and DREG series), and results with these monoclonal antibodies have been generally consistent with those found using the previously identified monoclonals.[34-37] One notable difference, however, appears to be the expression of the homing receptor on thymocyte populations of mouse and humans. In the case of murine thymocytes, the mature thymocyte population appears to specifically express the homing receptor (i.e., are MEL-14 positive),[16] while in humans, both mature and immature populations express the antigen.[31] While much of this work is phenomenologic, it is consistent with a potential role for this receptor in directing different cell types to different lymphoid and inflammatory tissue compartments (see below).

As has been reviewed elsewhere,[5,6,38] the homing receptor was found to be a member of a family of adhesion molecules that all seemed to utilize protein–carbohydrate interactions to mediate adhesion, and has been variously termed the *LECCAM* or *selectin* family. The second member of the family, the endothelial leukocyte adhesion molecule (ELAM), was found to have a very similar overall mosaic construction as the homing receptor, with the exception that this glycoprotein had six complement binding-like repeats. This molecule was found on the endothelium, bound all leukocytes to varying degrees, and was inducible by inflammatory mediators such as interleukin-1 or tumor necrosis factor. The third family member, GMP-140 or platelet activation-dependent granule-external membrane protein (PADGEM), also had a similar structure with the exception that it had eight to nine complement binding-like repeats. In addition, there appeared to be a soluble form of this glycoprotein that was missing the transmembrane domain, probably removed by mRNA splicing. This glycoprotein was found in platelet alpha granules and the Weibel-Palade bodies of endothelial cells. In both cell types, the protein could be rapidly expressed on the cell surface by thrombin activation of platelets or the endothelium, where the adhesive character of the glycoprotein endowed cells expressing it with the ability to bind neutrophils and monocytes. The overall degrees of sequence homology between these glycoproteins was high in the lectin and egf domains (~65%–70%) and somewhat lower in the rest of the molecules (~40%). A ligand for both ELAM and PADGEM/GMP-140 appears to be the tetrasaccharide sialyl Lewis-X[39,40] (see also Chapter 2).

GENOMIC STRUCTURE AND CHROMOSOMAL LOCALIZATION OF THE HUMAN AND MURINE HOMING RECEPTOR GENES

The overall mosaic structure of the human[41] and murine homing[42] receptors was consistent with the assembly of these glycoproteins from gene segments encoding separate functional domains. This hypothesis was borne out by the analysis of the structures of the human and murine homing receptor genes. These analyses demonstrated that the structure/function motifs found in these glycoproteins were indeed encoded by separate exons in the genome. Thus, it was found that the signal sequence, the lectin domain, the egf-like domain, each complement binding-like motif, and the transmembrane domain were all encoded by discrete exonic sequences. In addition, the initiator methionine codon was encoded by a separate exon, and the cytoplasmic domain was encoded by two short exons in both species. Interestingly, the intronic interruption site or "phase" of each exon (i.e., the nucleotide site in a given triplet codon that is interrupted by the intron) was the same as that found for analogous coding exons in other genes. Thus, other genes containing lectin-like, egf-like, and complement binding-like exons are all interrupted by introns at identical sites in the triplet codons, irregardless of what type of coding exon comes before or after the motif.[42,43] This finding is interesting in an evolutionary sense because it is consistent with the hypothesis that the homing receptor-encoding exons arose from primeval progenitor-type exons that were dispersed throughout the genome by exon shuffling followed by mutation and selection for certain functional attributes.

The high degree of sequence relatedness between the members of the LECCAM family was found to correlate well with their positions in the murine and human genome. By a series of chromosomal mapping studies, it was found that both the human and murine homing receptors mapped to syntenic regions of human and mouse chromosome 1.[41–44] This data was made even more interesting by the finding that the ELAM and GMP-140/

PADGEM genes also mapped to this same region in both human and mouse genomes. Pulse-field gel analysis demonstrated that these three genes were within approximately 200,000 bases of each other.[44] These data were consistent with the derivation of these genes by amplification of a single progenitor LECCAM/selectin gene whose progeny ultimately produced the three members of this family by mutation and selection. It is also interesting to note that although these genes physically map very close to one another, their regulation is highly specialized, so that each gene product is generated in dramatically different ways. This suggests that divergent gene regulatory pathways co-evolved with the individual LECCAM/selectin genes to allow for a greater level of regulation with respect to tissue type and temporal expression.

THE NATURE OF THE LIGANDS FOR THE HOMING RECEPTOR

From the data cited thus far, it seemed clear that the interaction between the homing receptor and its ligand was, at least in part, due to the recognition of a carbohydrate(s) by the lectin domain. Thus, the blocking of cell adhesion by carbohydrates, the calcium dependence of endothelial binding, and the abolition of such binding by sialidase combined with the calcium-dependent lectin homology discovered in the homing receptor glycoprotein sequence were all strongly consistent with this possibility.[11] In addition, it was shown that the isolated natural receptor could interact with carbohydrates such as PPME, further strengthening the notion that this receptor was, in fact, a carbohydrate-binding protein or lectin.[45,46] A further piece of evidence came when the epitope for the MEL-14 adhesion-blocking antibody was mapped. These data were consistent with previous data regarding the blocking of PPME binding by MEL-14 in that they showed that this antibody mapped to the N-terminus of the lectin (potential carbohydrate-binding) domain.[47] The mapping data, together with the adhesion-blocking ability of the antibody, were also consistent with a role for the lectin domain in mediating cell–endothelial adhesion by recognition of a specific carbohydrate(s) located on the endothelial cell surface.

In order to analyze the nature of the ligand(s) for the homing receptor, a reagent specific for this molecule had to be produced. Streeter and colleagues took a traditional approach to this problem by generating murine monoclonal antibodies directed against antigens specific for peripheral lymph node high endothelial venules.[48] One such antibody, termed *MECA 79,* specifically recognized an antigen(s) on this type of endothelium and also blocked the ability of lymphocytes to bind to these endothelial cells in frozen section assays. Analysis of the glycoproteins recognized by this monoclonal antibody demonstrated a relatively broad specificity, with a number of surface glycoproteins reacting with the antibody. This result, coupled with the relatively low affinity and IgM isotype of the antibody, suggested that this monoclonal antibody may have been recognizing a carbohydrate epitope potentially involved in homing receptor–ligand interactions. A second, much more novel technology was utilized by Watson and colleagues to analyze the nature of the ligand(s) recognized by the homing receptor.[49] In this procedure, the extracellular domain of the murine receptor was ligated to the hinge, C_H2, and C_H3 motifs of the human IgG_1 molecule. This resulted in an artificial antibody-like molecule, termed the *homing receptor IgG chimera,* whose specificity for antigen was determined by the ability of the homing receptor to adhere to its ligand(s). In addition, the IgG region could be utilized for a number of purposes, including: 1) dimerization by virtue of the disulfide-bonded hinge region leading to enhanced avidity, 2) ease of purification due to the specific interaction of this region with protein A or G, and 3) the ability to analyze receptor–ligand interactions by immunohistochemistry or immunoprecipitation using readily available reagents. This reagent was found to recognize carbohydrates such as PPME, inhibit lymphocyte binding to endothelial cells in frozen section assays, and, perhaps most interestingly, to specifically stain peripheral lymph node high endothelial venules but not Peyer's patch venules. This staining was calcium-dependent, MEL-14 inhibitable, and blocked by carbohydrates such as fucoidin, consistent with a protein–carbohydrate interaction. These results demonstrated that this reagent could be utilized for histochemical studies to localize the ligand(s) for the homing receptor. In addition, it was consistent with the hypothesis that specific trafficking of lymphocytes to peripheral lymph nodes via the homing receptor-dependent pathway was due to the lack of ligand expression in the Peyer's patch. This exciting result suggested that region-specific ligand expression was indeed one of the mechanisms that directed specific leukocyte trafficking to various lymphoid sites.

While the above experiments suggested the usefulness of the homing receptor–IgG chimera as a reagent for analyzing the tissue distribution of the ligand(s) for the homing receptor, they only began to describe the biochemical nature of this ligand(s). In order to examine the potential carbohydrate nature of the ligand(s), tissue sections were treated with sialidase and were then stained with the homing receptor–IgG chimera. These results were consistent with previous data[21a,51] and demonstrated that removal of sialic acid resulted in an abolition of staining, again suggesting that the homing receptor–ligand interaction involved, in part, sialic acid recognition.[52–54] Imai and colleagues[55] carried the analysis of the ligand(s) much further by taking advantage of an interesting, although unexplained, observation made many years earlier. This early work demonstrated that high endothelial venules of peripheral lymph nodes rapidly incorporated large amounts of inorganic sulfate into a secreted macromolecule.[56] At the time, these investigators proposed that this label was incorporated into a secreted glycolipid. Interestingly, autoradiographic analysis demonstrated that sulfate incorporation was found over the Golgi complex at early labelling times, also consistent with a potential glycoprotein localization of the label. Imai et al. reasoned that the tissue-specific localization of the sulfate labelling in addition to its localization over the endothelial venules suggested that this atom may have been incorporated into a ligand(s) for the homing receptor. Immunoprecipitation analysis of sulfate-labelled murine peripheral lymph nodes using the homing receptor IgG chimera revealed that a specific ~50-kD glycoprotein appeared to interact with this chimera. In addition, a ~90-kD protein was also immunoprecipitated by the receptor chimera, although it was labelled much less intensely. Surprisingly, analysis of the total sulfate-labelled material revealed that, in a short (~2 hours) pulse label, the ~50-kD protein band appeared to be the major sulfate-labelled molecule. In addition, and in agreement with the immunohistochemical results, the synthesis of these ligands was highly tissue-specific, with no ligand detected in other lymphoid tissues such as spleen or Peyer's patch. The interaction of both the 50-kD and 90-kD putative ligands with the homing receptor chimera was calcium-dependent and blocked by the MEL-14 antibody as well as by certain carbohydrates such as fucoidin and PPME. These results were consistent with the possibility that the interaction between the homing receptor and its ligand was, at least in part, due to protein (i.e., lectin domain)–carbohydrate interactions. This notion was further supported by experiments demonstrating that treatment of the isolated 50-kD protein with sialidase resulted in a ~5-kD decrease in molecular weight and a concomitant loss in ability to interact with the homing receptor chimera. This result was entirely consistent with all previous data concerning the role of sialic acid in adhesion mediated by the homing receptor. The carbohydrate attached to these ligands appeared to be all O-linked,[57] as treatment with N-glycanase showed no diminution of molecular weight. Referring to previous work by Streeter et al.,[48] analysis of the reactivity of the isolated ~50-kD ligand with MECA 79 revealed that this glycoprotein effectively interacted with this antibody, confirming previous data suggesting that this antibody recognized a component of the ligand for the homing receptor. Finally, protease digestion of the isolated ~50-kD band revealed that this component was a glycoprotein. In conclusion, this work revealed that the ligand(s) for the homing receptor were two sulfate-labelled, peripheral lymph node-specific glycoproteins whose interaction with this adhesive glycoprotein were in part mediated by carbohydrates, especially sialic acid. The relevance of the O-linkage of the sugar is presently not understood but may have to do with an enhanced presentation of the relevant carbohydrate residues to the homing receptor lectin domain.[57] A similar, although much less clear cut, analysis of the homing receptor ligand(s) has recently been presented.[58]

STRUCTURE AND FUNCTION OF HOMING RECEPTOR DOMAINS

While the data concerning calcium dependence, carbohydrate and antibody blocking, and sialidase sensitivity all argued strongly for a role for the putative carbohydrate-binding or lectin domain in adhesion mediated by the homing receptor, it is possible that other domains are involved with cell adhesion as well. An interesting phenomenon was noted when it was reported that recognition of the lectin-localized MEL-14 epitope appeared to be dependent upon the inclusion of the egf-like domain.[47] This phenomenon has also been noted for the ELAM molecule as well as for the human homologue of the homing receptor (Tedder,T., personnal communication). Assuming that the epitopes for these various antibodies are located exclusively within the lectin domain, these results are consistent with the interpretation that the egf domain has an effect on the overall conformation of the lectin domain. In addition, the work of Siegelman et al.[50] has suggested a potentially even more important role for the egf-like domain in cell adhesion. These authors demonstrated that the epitope recognized by the allotypic antibody LY22 was located in the egf domain of the homing receptor, where a single amino acid polymorphism was found to be responsible for the generation of this allotypic response. These authors went on to show that this antibody could effectively block the binding of lymphocytes to peripheral lymph node endothelium, suggesting that the egf-like domain may be directly involved with cell adhesion. In addition, they showed that the binding of this antibody enhanced the binding of the carbohydrate PPME to the homing receptor. Assuming that this antibody actually binds exclusively to the egf-like domain, this enhancement is consistent with a modification of the conformation of the lectin domain by antibody-induced perturbation of the egf domain. Thus, these results may be interpreted to agree with previous data indicating a role for the egf domain in lectin domain structure, as well as suggesting a more direct role for the egf-like domain in cell adhesion. It must be cautioned, however, that the enhancement of binding to PPME, an artificial binding substrate, cannot be taken to mean a necessary enhancement of binding to the natural ligand, so that this apparent modification of the lectin domain by antibody binding may have had an adverse effect on recognition of the endothelial ligand. It is therefore conceivable that a conformational modification of the lectin domain may have been responsible for the inhibition of adhesion by the LY22 antibody. The direct role of the egf-like domain in cell adhesion, therefore, awaits further experimentation.

While these data suggested several potential roles for the lectin and egf domains of the homing receptor, the function(s) of the complement binding-like motifs were still not understood. The somewhat lower degree of homology between the human and murine forms of these domains compared to the lectin and egf-like motifs suggested a less stringent role for these regions in homing receptor function. The identity of the two complement-binding motifs in the murine homing receptor at the nucleotide level suggested a critical need for two of these repeats, although the fact that the human motifs are not identical to one another suggests that they need to have homology but not identity for appropriate function. In order to analyze the role(s), if any, of these motifs in lectin domain function, homing receptor-IgG chimeras were produced with deletions of these duplicated domains.[59] The first surprising results obtained with this deletion mutant was that reactivity with the MEL-14 antibody dropped precipitously, suggesting a role for these motifs in lectin domain structure potentially analogous to that suggested for the egf domain. Not surprisingly, this loss of MEL-14 reactivity was accompanied by a loss of carbohydrate (PPME)-binding capacity, suggesting that the diminution of MEL-14 binding activity was probably due to a perturbation of the lectin domain conformation. This mutant was also unable to recognize the endothelial ligand for the homing receptor in cell blocking, immunohistochemical staining, and immunoprecipitation experiments. These results were all consistent with a role for the complement-binding repeats in appropriate lectin domain function. While a direct conformational interaction between the lectin and complement-binding motifs may be one potential explanation for this finding, a more likely one is that these domains serve to form oligomeric complexes that increase avidity as well as serve to induce appropriate lectin conformation. Other type C lectins appear to show a preference for oligomer formation, and at least one other selectin, ELAM, appears to form oligomeric complexes when expressed as a secreted molecule.[60] The individual domains of the homing receptor appear to act in complex coordinated manners that are still not fully understood.

THE NEUTROPHIL HOMING RECEPTOR: A MEDIATOR OF NEUTROPHIL ROLLING *IN VIVO*

While it seemed clear that the function of the homing receptor of lymphocytic cells was to enable them to efficiently traffic to peripheral lymphoid organs, the location of this receptor on other leukocytes, such as neutrophils,[61] was somewhat less clear. Cells such as neutrophils do not ordinarily traffic to peripheral lymph nodes but instead are found in acute inflammatory sites Thus, it was thought that the neutrophil homing receptor might function as a mediator of neutrophil adhesion during some step of acute inflammation. This possibility was supported by early experiments that demonstrated that acute neutrophil-mediated inflammation in a number of *in vivo* inflammatory models could be blocked by the MEL-14 antibody. Later experiments revealed that removal of the homing receptor from the lymphocyte surface by either activation (see below) or by mild trypsin digestion resulted in cells that were unable to appropriately migrate to inflammatory sites,[62,63] again consistent with a role for this adhesion molecule in inflammation. While these results implied a function for the homing receptor in neutrophil inflammatory processes, the experiments were criticized because either the mere binding of the antibody to the neutrophil surface might have had deleterious effects on neutrophil function that were not related to adhesion or the activation and/or proteolysis of the neutrophil may have had more pleiotropic effects than mere removal of the homing receptor. Another approach was taken by Watson et al.[64] when they examined the ability of the previously described homing receptor–IgG chimera to inhibit acute neutrophil-mediated inflammation in the peritoneal inflammatory model. In this model, inoculation of thioglycollate into the mouse peritoneum results in a rapid and profound neutrophil influx. These investigators found that intravenous administration of ~30 microgram per ml of the homing receptor chimera to mice before the induction of the peritoneal inflammatory response resulted in a profound inhibition of the ability of neutrophils to traffic to the peritoneum. The chimera also resulted in a significant, but lesser, degree of inhibition of the trafficking of lymphocytes to peripheral and mesenteric, but not to Peyer's patch, lymph nodes. As little as ~3 micrograms per ml gave a significant inhibition of neutrophil influx. In addition, these investigators found that the inhibition was almost complete at two hours after the induction, but appeared to be less significant at four hours. The interpretation of this work was that the soluble adhesion chimera competitively inhibited the adhesive interaction between the neutrophil homing receptor and its ligand(s) on the endothelial cell surface adjacent to the inflammatory site. These results were consistent with a significant role for the neutrophil homing receptor in acute inflammatory responses, and also implied that a carbohydrate-like ligand similar to that found in the peripheral lymph node may have been responsible for neutrophil trafficking to acute inflammatory sites.

The above data, while indicating a critical role for the homing receptor in neutrophil-mediated inflammation, did not directly identify what this specific task(s) was. For example, the homing receptor may have been involved in the efficient high avidity contact between neutrophils and the endothelium in a manner similar to that proposed for the interaction(s) between the lymphocyte and the postcapillary venule of the peripheral lymph node.[65] Alternatively, it was possible that this adhesion molecule was involved in a physiological function first identified about 100 years ago and termed *neutrophil rolling*. Using the so-called intravital microscopy technique, examination of postcapillary venules during the early stages of acute inflammation revealed that neutrophils interacted with the endothelium in a low affinity manner, such that the cells were observed to roll along the endothelium at a rate much slower than the rate seen for the unattached cells in the circulation.[66,67] This rolling episode was found to be the precursor for the later higher affinity events that resulted in the complete arrest of neutrophil motion followed by a change in the shape of the cell from a rounded to flattened morphology and finally by diapedesis across the endothelial barrier into the inflammatory site. Thus, it could be concluded that neutrophil, and perhaps other leukocytic cell, inflammation appeared to be a multistep phenomenon consisting of initial low-affinity interactions followed by higher-affinity adhesive events.

In order to examine the possibility that the neutrophil homing receptor was involved in leukocyte rolling, the same technique that was used to originally describe this phenomenon was applied, using more modern techniques and reagents. In these studies, Ley et al.[68] utilized intravital videomicroscopy of rat mesenteric postcapillary venules to examine the effects of the homing receptor chimera and a polyclonal antibody directed against the homing receptor on neutrophil rolling *in vivo*. An adjacent upstream venule was cannulated and injected with various solutions, then videomicroscopy of the downstream venule was utilized to quantitate the numbers of neutrophils rolling during a given time period. These investigators found that injection of a 100 microgram per ml solution of the murine homing receptor chimera inhibited neutrophil rolling by ~85%, while injection of a human CD4-IgG chimera gave no such inhibition. In addition, injection of a polyclonal rabbit antibody directed against the murine homing receptor also inhibited neutrophil rolling by ~80%, while a pre-immune serum showed no such effect. The authors also examined the temporal effects of inhibitor administration by investigating the rate that normal leukocyte rolling recurred after the termination of a constant perfusion of either the homing receptor chimera or polyclonal anti-homing receptor antibody. These studies showed recurrence of normal neutrophil rolling approximately 15 to 20 seconds after termination of inhibitor perfusion. In the case of the chimera, these results were consistent with a low-affinity interaction between the homing receptor chimera inhibitor and the endothelial ligands(s) for this adhesion molecule, so that the inhibitor dissociated from the ligand relatively quickly. This type of low-affinity interaction would be exactly as expected for the rolling phenomenon, where adhesive interactions would be expected to be made and broken with quite regular frequency. These elegant experiments were thus consistent with a critical function for the neutrophil homing receptor in leukocyte rolling near sites of chronic inflammation. In addition, they served to explain the *in vivo* blocking results of Watson et al.[64] and suggested that inhibition of neutrophil rolling is accompanied by a concomitant inhibition of tissue inflammation. In conclusion, these studies demonstrated that neutrophil rolling is, at least in part, mediated by the homing receptor and that the rolling phenomenon is a critical precursor to other adhesive and migratory aspects of the inflammatory response.

REGULATORY ASPECTS OF HOMING RECEPTOR FUNCTION

The regulation of homing receptor function is apparently accomplished using a number of different mechanisms. Perhaps the most basic is the regulation of cell type and developmental time of the expression of this adhesive glycoprotein.[31,33,36] In the case of the lymphocytic population, expression of the homing receptor appears to await entry of these cells from the bone marrow into the thymic compartment. The antigen is then expressed during thymic development so that lymphocytes released from this compartment appropriately traffic to peripheral and mesenteric lymph nodes. It may be assumed that expression of the glycoprotein before thymic education would result in trafficking of unselected bone marrow lymphocytes directly to peripheral lymphoid compartments, with the possibility that auto-immune like syndromes would develop. Thus, the highly regulated expression of this adhesion molecule in these cells appears to insure appropriate thymic education before release of the cells to the periphery. In contrast to this situation, expression on myeloid cells appears to occur much earlier in bone marrow development. In humans, it has been shown that expression occurs on very early myeloid precursor cells, suggesting other potential roles for this molecule in adhesion and trafficking. The early expression on myeloid cells is expected in view of the fact that these leukocytic cells are released directly into the circulation so that they can function immediately. In addition, the high turnover rate of the granulocyte population is consistent with a rapid release of neutrophils whose functional capacity to invade inflammatory sites is pre-existent. Thus, it may be concluded that regulation of expression of the homing receptor on various cell types during hematopoiesis is a key step in appropriate function of these cells during peripheral trafficking.

A second interesting aspect of homing receptor regulation concerns the ability of cells to rapidly shed the glycoprotein from their surfaces.[35,36,69-76] In the case of neutrophils, this shedding appears to occur within seconds of a number of physiologic stimuli, many of which appear to be involved with neutrophil activation. The release from lymphocytes, especially in response to antigenic stimuli, appears to occur with much slower kinetics. Interestingly, the regulation of release of the homing receptor from the neutrophil surface is a mirror image of the activation of the adhesive integrins of the beta 2 type (i.e., CD11/CD18). Thus, loss of the homing receptor after neutrophil activation is accompanied by a concomitant upregulation of expression and presumed adhesive capacity mediated by the beta 2 integrins. This regulation, when viewed in light of previous data concerning the function of the homing receptor in neutrophil rolling, suggests a model where rolling neutrophils, once activated by factors near the inflammatory site, lose their surface homing receptors and bind with high affinity to the endothelium by virtue of the CD11/CD18 integrin complex. The loss of the surface homing receptor might insure that inappropriate release of the neutrophil from the endothelium would not result in activated, and potentially damaging, neutrophils from rolling and attaching to other noninflammatory sites.

The mechanism by which this rapid loss occurs is quite interesting. The slightly smaller size of the released material suggests that the shedding appears to be mediated by proteolysis at a site very close to the transmembrane anchor site of the receptor. The work of Camerini et al.[28] initially suggested that there may be two forms of the receptor, one with the accepted transmembrane anchor and a second with a glycophospholipid (GPI) anchor. This second form was hypothesized based upon a second cDNA clone with a potential GPI anchor signal (hydrophobic C-terminal domain) and a very limited transient expression experiment that seemed to demonstrate a lipid-anchored form of the molecule. However, a variety of data, including genomic structural analysis[41,42] that is consistent with a lack of alternative RNA splicing, lack of phospholipase C cleavage of the receptor on a number of cells, normal expression of the receptor on patients with paroxysmal nocturnal hemoglobinuria (PNH—a somatic mutation that prevents GPI linkage), and the absence of any other mRNAs by polymerase chain reaction analysis,[41] argue that the second message seen by Camerini et al. may have been a cDNA cloning artifact. The nature of the protease that mediates this cleavage may be unique. The high degree of conservation of the transmembrane anchor regions between the human and murine receptors[27] suggests that the protease may be a membrane-anchored one that adheres to the homing receptor specifically through this conserved domain. The tight linkage of the protease to the receptor is supported by data demonstrating that inclusion of a variety of protease inhibitors in the media of cells expressing the receptor does little to inhibit the shedding of the molecule. Its potential linkage with cell signalling elements is supported by the finding that agonists of protein kinase C activity appear to induce rapid cleavage of the receptor from the cell surface, while inhibitors of this kinase appear to decrease the rate of shedding.[71,74] Thus, the proteolytic cleavage of this adhesion receptor and resultant shedding appear to be highly regulated functions.

A third important aspect of homing receptor regulation revolves around an apparent change in the avidity of the receptor for carbohydrate and endothelial ligand binding. The work of Spertini et al.[77] revealed that appropriate activation of lymphocytes or other cells resulted in an increased affinity of the homing receptor for the carbohydrate, PPME. This enhanced affinity was accompanied by an increased binding of the activated cells to the endothelium of peripheral lymphoid tissue in the frozen section assay. These results may be interpreted in the context of other adhesion molecules, such as the beta 2 integrins, which also appear to show an enhanced avidity for their ligand(s) after cell activation. In the case of the integrins, and presumably the homing receptor as well, the enhanced affinity is due to a conformational change in the protein. This conformational change may be in the overall three-dimensional structure of the protein or it may be due to oligomerization of the receptors with a resultant enhanced binding avidity. This level of regulation thus allows for a cell type-specific activation of the adhesion mediated by this receptor, as the enhanced avidity would be dependent upon the type of activation and the spectrum of receptors (i.e., antigen-specific, chemotactic) found on the cell surface. This type of activation would thus partially explain why only specific types of cells appear to traffic to various sites, in spite of the fact that a number of nontrafficking cells also possess cell surface homing receptor. Thus, the specific trafficking of lymphocytes to peripheral lymphoid tissue may be accomplished by lymphocyte-specific activation signals in these organs, while the rapid and specific early trafficking of neutrophils to sites of acute inflammation may be accomplished by a divergent set of neutrophil-specific activators produced at these acute inflammatory sites. The possibility that this activation event is dependent upon protein kinase C-mediated phosphorylation of the cytoplasmic domain has been previously alluded to and awaits experimental confirmation.

A final, although unproven, level of regulation may involve signalling mediated by the homing receptor to activate cell motility and/or shape. A potentially significant set of data in this regard suggests that exposure of lymphocytes to pertussis toxin, an inhibitor of G protein-mediated cell signalling, appears to profoundly affect lymphocyte trafficking to peripheral lymphoid organs.[78] Indeed, animals given pertussis toxin show a high level of lymphocytosis, suggesting that these cells are incapable of migrating through the peripheral lymphoid organ pathway, thus resulting in an increase in the levels of circulating lymphocytes. While these results may be interpreted in a number of different ways, one possible interpretation is that the homing receptor interacts with one or more G proteins on the cytoplasmic side, perhaps through protein–protein interactions in the highly conserved transmembrane domain region, so that contact of the homing receptor with the endothelium activates a cellular signalling cascade that results in changes in cytoskeletal elements and resultant cell motility into the lymphoid organ. While the veracity of this proposal remains to be determined, one report suggests that an antibody directed against the human homing receptor can inhibit B-cell differentiation, consistent with a potential signalling mechanism of B cell homing receptor.[79]

SUMMARY

The homing receptor is the paradigm for a family of inflammatory adhesion molecules that mediate leukocyte–endothelial binding through protein–carbohydrate interactions. The homing receptor appears to have a critical function in normal peripheral lymph node trafficking as well as in acute inflammatory responses mediated by neutrophils. It remains to be seen whether this adhesion molecule is involved in other inflammatory conditions, particularly chronic inflammatory syndromes such as arthritis or autoimmune disease. The involvement of this adhesion receptor in acute neutrophil-mediated syndromes suggests potential clinical uses for inhibitors directed against the adhesive functions of the homing receptor. Such inhibitors, including soluble receptor–immunoglobulin chimeras such as those described previously or carbohydrate-like molecules based upon the naturally occuring ligands,[80] will undoubtedly be clinically tested in the very near future. It is hoped that the knowledge produced from basic investigations into the molecular and cellular biology of the homing receptor may ultimately result in efficient antiinflammatory compounds for a number of currently untreatable clinical conditions.

REFERENCES

1. Anderson, D. C., and Springer, T. A. (1987) *Annu. Rev. Med.* 38, 175–194.
2. Arnaout, M.A. (1990) *Blood* 75, 1037–1050.
3. Carlos, T.M. and Harlan, J.M. (1990) *Immunol. Rev.* 114, 5–28.
4. Springer, T.A. (1990) *Nature* 346, 425–434.
4a. Arfors, K.E., Lundberg, C., Lindbom, L., Lundberg, K., Beatty, P.G., and Harlan, J.M. (1987) *Blood* 69, 338–342.
5. Lasky, L.A. (1991) *J. Cell. Biochem.* 45, 139–146.
6. Lasky, L.A. and Rosen, S.D. (1991) The Selectins: Carbohydrate-Binding Adhesion Molecules of the Immune System. In *Inflammation: Basic Principles and Clinical Correlates,* (Gallin, J., Goldstein, I., and Snyderman, R., eds.) Raven Press. In press.
7. Berg, E.L., Goldstein, L.A., Jutila, M.A., Nakache, M., Picker, L.J., Streeter, P.R., Wu, N.W., Zhou, D., and Butcher, E.C. (1989) *Immunol. Rev.* 108, 1–18.
8. Duijvestijn, A. and Hamann, A. (1989) *Immunol. Today* 10, 23–28.
9. Ford, W.L. (1969) *Cell Tissue Kinet.* 2, 171.
10. Gowans, J.L. (1959) *J. Physiol.* 146, 54.
11. Rosen, S.D. (1989) *Curr. Opin. Cell. Biol.* 1, 913–919.
12. Stoolman, L.M. (1989) *Cell* 56, 907–910.
13. Yednock, T.A. and Rosen, S.D. (1989) *Adv. Immunol.* 44, 313–378.
14. Stamper, H.B. and Woodruff, J.J. (1976) *J. Exp. Med.* 144, 828–833.
15. Woodruff, J.J., Clarke, L.M., and Chin, Y.H. (1987) *Ann. Rev. Immunol.* 5, 201–222.
16. Gallatin, M., St. John, T., Siegelman, M., Reichert, R., Butcher, E., and Weissman, I. (1986) *Cell* 44, 673–680.
17. Gallatin, W.M., Weissman, I.L., and Butcher, E.C. (1983) *Nature* 303, 30–34.
18. Stoolman, L.M. and Rosen, S.D. (1983) *J. Cell Biol.* 96, 722–729.
19. Stoolman, L.M., Tenforde, T.S., and Rosen, S.D. (1984) *J. Cell Biol.* 99, 1535–1540.
20. Stoolman, L.M., Yednock, T.A., and Rosen, S.D. (1987) *Blood* 70, 1842–1850.

21. Yednock, T.A., Stoolman, L.M., and Rosen, S.D. (1987) *J. Cell Biol.* 104, 713–723.
21a. Rosen, S.D., Chi, S.I., True, D.D., Singer, M.S., and Yednock, T.A. (1989) *J. Immunol.* 142, 1895–1902.
22. Yednock, T.A., Butcher, E.C., Stoolman, L.M., and Rosen, S.D. (1987) *J. Cell Biol.* 104, 725–731.
23. Lasky, L.A., Singer, M.S., Yednock, T.A., Dowbenko, D., Fennie, C., Rodriguez, H., Nguyen, T., Stachel, S., and Rosen, S.D. (1989) *Cell* 56, 1045–1055.
24. Siegelman, M.H., Van de Rijn, M., and Weissman, I.L., (1989) *Science* 243, 1165–1172
25. Drickamer, K. (1988) *J. Biol. Chem.* 263, 9557–9560.
26. Reid, K. and Day, A.J. (1989) *Immunol. Today* 10, 177–180.
27. Bowen, B.R., Nguyen, T., and Lasky, L.A. (1989) *J. Cell Biol.* 109, 421–427.
28. Camerini, D., James, S.P., Stamenkovic, I., and Seed, B. (1989) *Nature* 342, 78–82.
29. Siegelman, M.H. and Weissman, I.L. (1989) *Proc. Natl. Acad. Sci. USA* 86, 5562–5566.
30. Tedder, T.F., Isaacs, C.M., Ernst, T.J., Demetri, G.D., Adler, D.A., and Disteche, C.M. (1989) *J. Exp. Med.* 170, 123–133.
31. Tedder, T.F., Penta, A.C., Levine, H.B., and Freedman, A.S. (1990) *J. Immunol.* 144, 532–540.
32. Kansas, G.S., Muirhead, M.J., Dailey M.O. (1990) *Blood* 76, 2483–2492.
33. Picker, L.J., Terstappen, L.W.M.M., Rotts, L.S., Streeter, P.R., Stein H., and Butcher E.C. (1990) *J. Immunol.* 145, 3247–3255.
34. Jutila, M.A., Kishimoto, T.K., and Butcher E.C. (1990) *Blood* 76, 178–183.
35. Kishimoto, T.K., Jutila, M.A., Berg, E.L., and Butcher, E.C. (1989) *Science* 245, 1238–1241.
36. Kishimoto, T.K., Jutila, M.A., and Butcher, E.C. (1990) *Proc. Natl. Acad. Sci. USA* 87, 2244–2248.
37. Stoolman, L.M. and Ebling, H. (1989) *J. Clin. Invest.* 84, 1196–1205.
38. Osborn, L. (1990) *Cell* 62, 3–6.
39. Brandley, B.K., Swiedler, S.J., and Robbins, P.W. (1990) *Cell* 63, 861–863.
40. Springer, T.A. and Lasky, L.A. (1991) *Nature* 349, 196–197.
41. Ord, D.C., Ernst, T.J., Zhou, L.J., Rambaldi, A., Spertini, O., Griffin, J., and Tedder, T.F. (1990) *J. Biol. Chem.* 265, 7760–7767.
42. Dowbenko, D., Diep, A., Taylor, B., Lusis, A., and Lasky, L. (1991) *Genomics* 9, 270–279.
43. Collins, T., Williams, A., Johnston, G.I., Kim, J., Eddy, R., Shows, T., Gimbrone, M.J., and Bevilacqua, M.P. (1991) *J. Biol. Chem.* 266, 2466–2473.
44. Watson, M.L., Kingsmore, S.F., Johnston, G.I., Siegelman, M.H., Le, B.M., Lemons, R.S., Bora, N.S., Howard, T.A., Weissman, I.L., McEver, R.P., and Seldin, M.F. (1990) *J. Exp. Med.* 172, 263–272.
45. Geoffroy, J.S. and Rosen, S.D. (1989) *J. Cell Biol.* 109, 2463–2469.
46. Imai, Y., True, D.D., Singer, M.S., and Rosen, S.D. (1990) *J. Cell Biol.* 111, 1225–1232.
47. Bowen, B., Fennie, C., and Lasky, L.A. (1990) *J. Cell Biol.* 110, 147–153.
48. Streeter, P.R., Rouse, B.T.N., and Butcher, E.C. (1988) *J. Cell Biol.* 107, 1853–1862.
49. Watson, S.R., Imai, Y., Fennie, C., Geoffroy, J.S., Rosen, S.D., and Lasky, L.A. (1990) *J. Cell Biol.* 110, 2221–2229.

50. Siegelman, M.H., Cheng, I.C., Weissman, I.L., and Wakeland, E.K. (1990) *Cell* 61, 611–622.
51. Rosen, S.D., Singer, M.S., Yednock, T.A., and Stoolman, L.M. (1985) *Science* 228, 1005–1007.
52. Schauer, R. (1982) *Adv. Carbohydr. Chem. Biochem.* 40, 131–233.
53. Schauer, R. (1985) *Trends Biochem Sci.* 10, 357–360.
54. True, D.D., Singer, M.S., Lasky, L.A., and Rosen, S.D. (1990) *J. Cell Biol.* 111, 2757–2764.
55. Imai, Y., Singer, M.S., Fennie, C., Lasky, L.A., and Rosen, S.D. (1991) *J. Cell Biol.* 113, 1213–1221.
56. Andrews, P., Milsom, D., and Ford, W. (1982) *J. Cell Sci.* 57, 277–292.
57. Jentoft, N. (1990) *Trends Biochem. Sci.* 15, 291–294.
58. Berg, E.L., Robinson, M.K., Warnock, R.A., and Butcher, E.C. (1991) *J. Cell Biol.* 114, 343–349.
59. Watson, S.R., Imai, Y., Fennie, C., Geoffrey, J., Singer, M. Rosen, S.D. and Lasky, L.A. (1991) *J. Cell Biol.* 115, 235–244.
60. Lobb, R.R., Chi-Rosso, G., Leone, D.R., Rosa, M.D., Bixler, S., Newman, B.M., Luhowskj, S., Benjamin C.D., Dougas, I.G., Goelz, S.E., Hession, C. and Chow, E.P. (1991) *J. Immunol.* 147, 124–129.
61. Jutila, M.A., Rott, L., Berg, E.L., and Butcher, E.C. (1989) *J. Immunol.* 143, 3318–3324.
62. Jutila, M.A., Kishimoto K.T., Finken, M. (1991) *Cell. Immunol.* 132, 201–214.
63. Lewinsohn, D.M., Bargatze, R.F., and Butcher, E.C. (1987) *J. Immunol.* 138, 4313–4321.
64. Watson, S.R., Fennie, C., and Lasky, L.A. (1991) *Nature* 349, 164–167.
65. Butcher, E.C. (1986) *Curr. Top. Microbiol. Immunol.* 128, 85–122.
66. Fiebig, E., Ley, K., and Arfors, K.E. (1991) *Int. J. Microcirc.* 10, 127–144.
67. Ley, K., Lundgren, E., Berger, E., and Arfors, K.E. (1989) *Blood* 73, 1324–1330.
68. Ley, K., Gaehtgens, P., Fennie, C., Singer, M.S., Lasky, L.A., and Rosen, S.D. (1991) *Blood* 77, 2553–2555.
69. Berg, M. and James, S.P. (1990) *Blood* 76, 2381–2388.
70. Bochner, B.S. and Sterbinsky, S.A. (1991) *J. Immunol.* 146, 2367–2373.
71. Buhrer, C., Berlin, C., Thiele, H.G., Hamann, A. (1990) *Immunology* 71, 442–448.
72. Griffin, J.D., Spertini, O., Ernst, T.J., Belvin, M.P., Levine, H.B., Kanakura, Y., and Tedder T.F. (1990) *J. Immunol.* 145, 576–584.
73. Huang, K., Beigi, M., and Daynes R.A. (1990) *Reg. Immunol.* 3, 103–111.
74. Jung, T.M. and Dailey, M.O. (1990) *J. Immunol.* 144, 3130–3136.
75. Smith, C.W., Kishimoto, T.K., Abbass, O., Hughes, B., Rothlein, R., McIntire, L.V., Butcher, E., and Anderson D.C. (1991) *J. Clin. Invest.* 87, 609–618.
76. Spertini, O., Freedman, A.S., Belvin, M.P., Penta, A.C., Griffin, J.D., and Tedder, T.F. (1991) *Leukemia* 5, 300–308.
77. Spertini, O., Kansas, G.S., Munro, J.M., Griffin, J.D., and Tedder, T.F. (1991) *Nature* 349, 691–694.
78. Steen, P.D., Ashwood, E.R., Huang, K., Daynes R.A., Chung, H., and Samlowski, W.E. (1990) *Cell. Immunol.* 131, 67–85.
79. Murakawa Y., Strober W., and James, S.P. (1991) *J. Immunol.* 146, 40–46.
80. Weston, S.A. and Parish, C.R. (1991) *J. Immunol.* 146, 4180–4186.

CHAPTER 4

Regulation of Myeloid Blood Cell–Endothelial Interaction by Cytokines

Mathew A. Vadas, Jennifer R. Gamble, William B. Smith

Inflammatory reactions have a major cellular component that arises by the localized migration of cells from blood to tissue. This migration is an orderly, orchestrated, and usually self-limited event. In chronic inflammation, the controls limiting this reaction appear to be absent. This chapter discusses how the process is regulated by cytokines.

Blood Cell Endothelial Interactions and Their Measurement

In the postcapillary venules where migration takes place, the cells normally marginate and are seen to roll along the endothelium. During migration they become arrested and appear adherent. *In vitro* assays for measuring leukocyte–endothelial cell interactions utilize a monolayer of endothelial cells to which leukocytes are added for between five and 30 minutes and washed; the number of white cells associated with endothelium are then quantitated. A variation on this static assay is to perform this test under conditions of shear stress simulating flow conditions in capillaries or venules. It can be seen that both assays measure the number of white cells tightly associated with endothelium and do not differentiate cells that adhere to endothelium from those that adhere and migrate. This distinction may be thought not to matter, as adherence is often held to be a necessary preamble to transmigration. However, there exist some forms of cellular adherence that are not associated with transmigration; for example, as is seen after activation of white cells. *In vivo*, white cell activation can result in vessel plugging as seen in septicemia or in multitrauma. By contrast, the type of adhesion that is associated with transmigration seldom results in intravessel pathology. Thus, considerable care has to be taken in interpreting data that reports on adhesion reactions.

Cytokines and Adhesion

Small peptide regulatory molecules that act at short distances are usually referred to as *cytokines*. One of the chief types of signals that stimulate adhesion, whether it is accompanied by transmigration or not, are cytokines. It is likely that the type, amount, and sequence of localized secretion of cytokines determines the onset and offset of inflammatory reactions. The chief cytokines that induce a pro-adhesive state in endothelium are tumor necrosis factor (TNF)-α and interleukin-1 (IL-1)β. Endotoxin or lipopolysaccharide (LPS) has a very similar effect to TNF-α and IL-1β, and the three are referred to as the "septic triad."

Similarly, TNF-α, to a lesser extent interleukin-8 (IL-8), and the hemopoietic growth factors such as granulocyte macrophage-CSF (GM-CSF) are the important stimuli that act on neutrophils and other leukocytes to alter their adhesion. The effect of IL-8 and some chemoattractants, such as the bacterial product, f-met-leu-phe (FMLP), are partially overlapping.

It is seen that while cytokines may be the chief internal regulators, bacterial products (LPS, FMLP) and perhaps other agents can have similar phenotypic effects.

ACTIVATION OF ENDOTHELIUM BY CYTOKINES

Resting endothelium lacks molecules that support adhesion—a highly desirable state that appears to reflect the relatively uninterrupted flow of blood cells *in vivo*. There is detectable expression of intercellular adhesion molecule-1 (ICAM-1)[1] and vascular cell adhesion molecule-1 (VCAM-1)[2] on resting endothelial cells passaged *in vitro,* but it is uncertain whether they are present or functional *in vivo*. The demonstration that ICAM expressed in lipid bilayers is unable, by itself, to arrest cells subject to shear forces likely to be encountered *in vivo*[3] supports the notion that the mere expression of this molecule is insufficient for adhesion.

Upon addition of cytokines IL-1β or TNF-α to endothelium, there is a characteristic increase in adhesiveness for neutrophils,[4,5] eosinophils,[6] lymphocytes,[7] monocytes,[5,8] as well as tumor cells such as melanoma and colon carcinoma[9] and a change in the expression of adhesion molecules. TNF-β (lymphotoxin) has similar effects to TNF-α but is less potent in several assays.[10]

Effect of Cytokines on Endothelial Adhesion Structures

ICAM-1. ICAM, an adhesion molecule that has five immunoglobulin (Ig)-like extracellular domains,[11] is constitutively expressed on endothelium.[1] After activation with IL-1, TNF, or interferon-γ (IFN-γ), its expression is upregulated, peaking at about 24 hours but maintaining increased expression for at least 72 hours. Until recently, the functional ligands for ICAM have been controversial, as blocking experiments with antibodies gave different results. The situation has been clarified by the discovery[12] that the first (or most N-terminal) Ig domain of ICAM is the ligand for lymphocyte-function antigen-1 (LFA-1) (or the $\alpha_L\beta_2$ integrin), an adhesion molecule on most leukocytes. The third domain, especially in the deglycosylated form, is able to bind Mac-1 (or $\alpha_M\beta_2$ integrin), whose presence is restricted to myeloid cells.[13] Thus, antibodies against different epitopes would be expected to have different functional effects. As mentioned before, the mere presence of ICAM is insufficient to support adhesion under the shear forces likely to be encountered *in vivo*.[3] It appears that molecules of the lectin-epidermal growth factor-complement binding protein-cell adhesion molecule (LECCAM) class need to be involved in order to arrest leukocytes from the circulation and for ICAM-mediated interactions to become effective (see below).

VCAM-1. VCAM-1 is an adhesion molecule with six extracellular Ig-like domains,[14] that is variably expressed on resting endothelium. The expression of VCAM-1 is induced by IL-4 as well as IL-1 and TNF but not by IFN-γ.[15-17] It peaks at about six hours and declines to low levels by 48 to 72 hours.[15] VCAM-1 is the ligand for the integrin very late antigen-4 (VLA-4) ($\alpha_4\beta_1$), which is expressed on monocytes, lymphocytes, and eosinophils but not neutrophils.[14,15,18]

Endothelial Leukocyte Adhesion Molecule-1 (ELAM-1) or LECCAM-2. The adhesion protein ELAM-1 has extracellular domains composed of lectin, epidermal growth factor, and complement binding protein-like domains[19] and is not expressed on resting multipassaged endothelial cells.[20] The expression of ELAM-1 is induced by IL-1 and TNF-α but not by IFN-γ, peaks at four to six hours, and largely disappears by 24 hours.[20] There is a low expression on endothelial cells that have recently been derived from umbilical cords.[21]

ELAM-1 binds neutrophils,[20] memory T lymphocytes,[22,23] eosinophils,[24] and monocytes.[5,8] The ligand for ELAM-1 is the carbohydrate sialyl Lewis-X that is expressed on a number of normal and malignant cells on both lipids[25] and proteins.[26] Recently, sialyl Lewis-A has also been implicated.[27] There is also indirect evidence that ELAM-1 may interact with another member of the LECCAM family, LECCAM-1 or MEL-14.[28]

The adhesion molecules involved in these interactions are summarized in Table 4-1.

Sequential Movement of Cells During Inflammation. During inflammation there is a sequential movement of cells from blood to tissues, with neutrophils predominating in the first 24 hours and monocytes and lymphocytes predominating thereafter. To some degree, the expression of adhesion molecules may explain this sequence of cell accumulation if transmigration and adhesion are indeed coupled (Figure 4-1 and Table 4-2).

It is seen that treatment of endothelial cells (EC) with TNF will result in the sequential increased expression of ELAM-1, VCAM-1, and ICAM-1 (see Figure 4-1), with ELAM-1 declining first. This may explain the initial transient migration of neutrophils followed by the more chronic movement of monocytes and lymphocytes. Memory T cells appear to be an exception to this rule, and using ELAM-1[22,23] they may potentially move into lesions very early.

INHIBITION OF ENDOTHELIAL ACTIVATION

It is clearly a matter of major concern if the endothelium expresses adhesion structures inappropriately. Cytokines that inhibit the expression of adhesion structures appear to provide a major mechanism of control.

Table 4-1. Receptor-ligand Pairs for Leukocyte-Endothelial Adhesion

Adhesion Molecule on EC	*"Receptor" on Myeloid Cells*
ELAM-1 (LECCAM-2)	Sialyl Le^x Sialyl Le^a LECCAM-1?
VCAM-1	VLA-4 ($\alpha_4\beta_1$)*
ICAM-1	Mac-1 $\alpha_M\beta_2$* LFA-1 $\alpha_L\beta_2$*
GMP-140 (LECCAM-3)	Sialyl Le^x Le^x?
Mannose-rich sugars	LECCAM-1*

*High and low affinity states have been demonstrated.

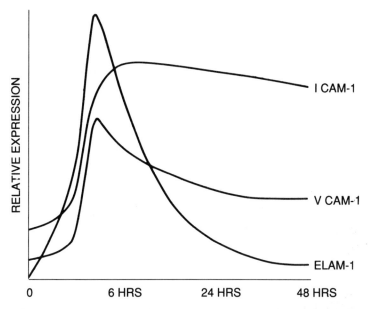

Figure 4-1. Schematic diagram of the expression of adhesion molecules on endothelial cells after stimulation with TNF-α or IL-1β. Vertical axis reflects expression on EC and horizontal axis time. Note that resting multi-passaged EC have no ELAM-1 and little VCAM-1. Also note that ELAM-1 expression appears to peak most quickly and to decline most rapidly.

Transforming Growth Factor-β (TGF-β)

TGF-β inhibits baseline adhesion and the IL-1 or TNF-mediated increase in adhesiveness for neutrophils or lymphocytes. TGF-β has to be present for at least nine hours before the addition of IL-1 or TNF and the potency of TGF-$β_1$ and TGF-$β_2$ appears the same.[29] This finding suggests that TGF-β

Table 4-2. Adhesion of Myeloid Cells to Endothelial Adhesion Proteins

	Neutrophils	*Eosinophils*	*Monocytes*	*Lymphocytes*
ELAM-1	+	+	+	−*
VCAM-1	−	+	+	+
ICAM-1	+	+	+	+

*except memory and skin homing lymphocytes
+ adhesion
− no adhesion

could provide a tonic inhibition of adhesion and thus, perhaps, inflammation *in vivo*. The full adhesive phenotype would need the presence of a pro-inflammatory cytokine and the absence of TGF-β.

Interestingly, the inhibition is seen only in recently-explanted EC from human umbilical vein, and the inhibitory effect of TGF-β on adhesion, but not other EC properties like cell division, is lost with passage. This phenomenon is paralleled by the inhibition of expression of ELAM-1 by TGF-β in recently-explanted or "young" but not *in vitro*-aged EC.[21] As noted previously, ELAM-1 expression was found in young EC that was not intentionally stimulated by IL-1 or TNF-α. The possibility thus arises that there exists two pathways of stimulating ELAM-1 expression, and one of these, the one that operates on "young" EC, is inhibitable by TGF-β. The relatively high amounts of immunoreactive TGF-β found around blood vessels may then serve a role in inhibiting activation of EC.[29a]

IL-4

IL-4 appears to have both a stimulatory and an inhibitory role on EC adhesion[30] It inhibits the expression of ELAM-1 and ICAM-1 but enhances the expression of VCAM-1 on endothelium.[31] Thus, the net effect of IL-4 on EC may favor the accumulation of lymphocytes rather than neutrophils into inflammatory sites. The finding that IL-4 also acts on monocytes to inhibit their adhesiveness[32] suggests that the chief effect may be to promote lymphocyte accumulation.

It is thus seen that the adhesive phenotype of EC will be determined by the composition of the mixture of cytokines to which it is exposed. This is not surprising, as antagonism and synergism are seen in most other cytokine-driven systems and allow for a gradation of effects and for binding in fail-safe systems.

ACTIVATION OF BLOOD CELLS BY CYTOKINES

Control of adhesion is further complicated by cytokines acting on leukocytes to alter their adhesive phenotype. In general, these actions are more rapid than the effects on EC and have a shorter time course. In contrast to the effects on EC that result in localized changes, there is little capacity to control the site or extent of effects on leukocytes in the circulation. Hence, more florid generalized pathology tends to result from the effect of cytokines on leukocytes. Probably the chief physiologic site of the action of cytokines on blood cells is in the tissues, and it is only under extreme conditions that they become exposed to large amounts in the circulation.

Neutrophils

Adhesion. The adhesion of neutrophils under nonshear conditions to "resting" or not intentionally stimulated endothelium can be increased by a range of cytokines, the most powerful of which is TNF-α. TNF stimulates neutrophils to temporarily increase their adhesiveness. The CD11b/CD18 molecule is absolutely essential for this type of adhesion and is likely to be directly mediated by it, primarily through a conformational change[33,34] rather than merely an upregulation of the number of molecules per neutrophil. The enhanced adhesiveness of neutrophils lasts 30 minutes to one hour[35] and appears to involve a decrease in the cyclic AMP levels.[36] It is theorized that the CD11b/CD18 is maintained in a nonadhesive state by high levels of cAMP, a state that is transiently reversed by TNF-α.

There is evidence that TNF-β and GM-CSF have effects qualitatively similar to TNF-α on the neutrophils; however, the potency of these molecules is considerably less. Interestingly, there exists considerable interperson variation in responsiveness to these cytokines; however, neutrophils that respond well to one cytokine also respond well to the others.[37] This suggests that at least for this response the cytokines may share a common final signal that shows considerable interperson variation.

The LECCAM-1 molecule on neutrophils is thought to be responsible for shear-resistant adhesion and for margination.[38] Some agents that activate neutrophils[39] or lymphocytes[40] increase the affinity of LECCAM-1 for at least one of its sugar ligands, phosphomannanester (PPME), by up to 47-fold.[39] This increase in affinity is very rapid at 37°C (within 5 minutes) and also takes place at 4°C but at a slower rate. This is the first example of LECCAM class of adhesion molecules undergoing altered affinity following activation. Immediately following this change in affinity the expression of LECCAM-1 is rapidly lost from the neutrophil surface, whether the activation was with chemotactic agents[41,42] or the cytokines TNF and GM-CSF.[40] These are the very agents that also upregulate the expression and function of CD11b/CD18 molecules responsible for shear-sensitive adhesion and permissive for transmigration.[42,43] This reciprocal change in adhesion structures is paradoxical. It has been hypothesized that LECCAM-1 and CD18-dependent steps are sequential in which LECCAM-1 arrests the neutrophil while CD18 assures its subsequent transmigration.[38] Interestingly, a similar synergistic relationship has been suggested for granule membrane protein-140 (GMP-140), expressed briefly on EC following activation, and ICAM-1 the counter receptor for CD11/CD18.[3] GMP-140 arrests flowing neutrophils, allowing ICAM-1 to bind and support their subsequent spreading. Granulocyte colony stimulating factor (G-CSF), a lineage-specific growth factor for neutrophils that has a variable effect on neutrophil function, fails to change the expression of LECCAM-1[44] and has no or a very weak proadhesive effect in static assays.

As TNF activation of neutrophils does not by itself cause transmigration (nor does it enhance and it may actually inhibit[45,46] the responsiveness to chemotactic stimuli), the possibility must arise that some forms of activation leading to increased adhesion (mainly CD18-dependent) prevent cellular movement. CD18, however, appears essential for transmigration, as a congenital absence of CD18 as seen in the disease, leukocyte adhesion deficiency (LAD), or antibodies against CD18 prevents transmigration.[47] Part of the explanation may be that TNF-activated neutrophils also undergo a membrane change that renders the cells stiffer,[48] hence, less ready to move in chemotactic gradients but obviously retaining the capacity to do so. This may also explain the tendency of activated neutrophils to plug lung microvessels. Antibodies against CD18 prevent an impressive range of neutrophil-mediated pathology,[49,50] central to all of which is the plugging of blood vessels. One possibility is that a degree of transmigration (e.g., insertion of a pseudopod) is necessary for the pathology to develop, and antibodies against CD18 prevent both adhesion and any capacity for migration.

Neutrophils and IL-8. IL-8 was originally isolated as a neutrophil chemotactic factor in the supernatant of activated mononuclear cells and was subsequently found to also be produced by fibroblasts, endothelial cells, and numerous other cell types. It was purified as a 72-amino acid peptide and has been cloned and sequenced.[51] Another form of IL-8, with an extra five N-terminal amino acids, was isolated as a leukocyte adhesion inhibitor from the supernatants of activated endothelial cells.[52,53] It was found in some assays to inhibit neutrophil adhesion (as was subsequently the 72-amino acid form), but only to activated EC.[54]

Further experiments on the effect of IL-8 on neutrophil adhesion to activated endothelium have, however, given conflicting results. The initial findings were a ten-minute chromium labelled neutrophil adhesion assay. By contrast, in a 30-minute assay with dye labelling, a consistent but small IL-8-stimulated increase in neutrophil adhesion to activated endothelium was found.[55] Both assays measure only endothelial-associated neutrophils and do not distinguish between those that were adherent to the upper surface of the endothelium and those that had transmigrated through the monolayer *in situ.* With a slide assay, which allows for this distinction to be made, the number of adherent IL-8 preincubated neutrophils was not significantly changed but virtually none transmigrated.[56] The overall result is a decrease in endothelium-associated neutrophils (Figure 4-2), which is not due to an effect on adhesion. The reason for the discrepancy in results of these three assay systems is at present unknown.

The effect of IL-8 on neutrophil adhesion to activated endothelium can also be examined from the point of view of the adhesion molecule/ligand pairs involved. Neutrophil CD18 has been shown to contribute to approximately half of this adhesion in static assays,[47] and the function of these

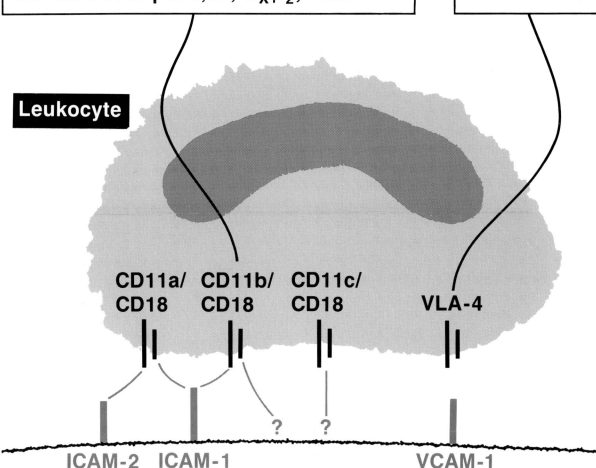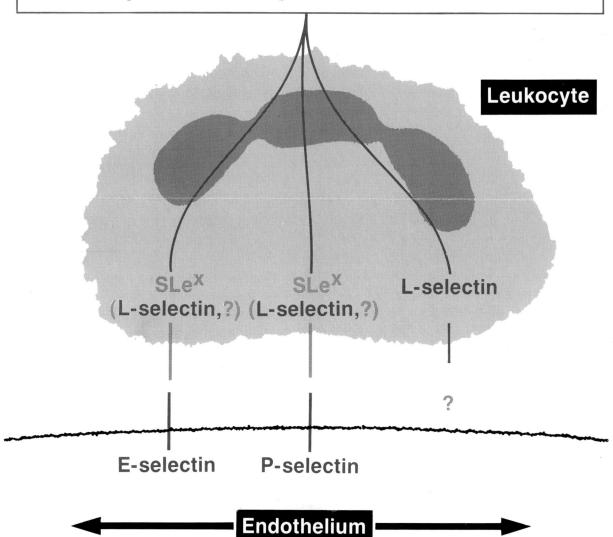

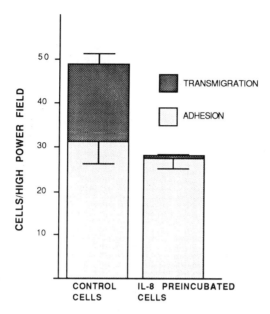

Figure 4-2. The interaction of IL-8 preincubated (100nM) and control preincubated neutrophils with IL-1 (100 U/ml) activated endothelium, considering transmigration and adhesion separately. Counts were made by phase–contrast microscopy. Total ECA (endothelial associated neutrophils) are shown by the total column heights.

molecules has been shown to be increased by IL-8 treatment.[57] Activated endothelium displays the adhesion molecule ELAM-1,[19,20] which binds neutrophils via the ligand sialylated Lewis-X,[25,58] the expression of which is unchanged on IL-8-treated neutrophils.[58]

IL-8-activated neutrophils also shed surface leukocyte endothelial cell adhesion molecule (LECAM-1), and such neutrophils should adhere less well to activated endothelium under conditions of flow, where CD18 mediated adhesion is not relevant.[43] The reason that adhesion of IL-8-activated neutrophils is preserved in some static assays despite loss of LECAM-1 may be that the increase in CD18 function compensates for this loss. GMP-140 may also compensate under appropriate circumstances[3] (see below). It is therefore difficult to completely account for the changes in adhesion mediated by IL-8 in terms of changes in adhesion molecules.

Treatment of neutrophils with exogenous IL-8 has other effects in addition to alterations in adhesion molecules, such as activation and calcium flux. Neutrophils can also become desensitized selectively to IL-8 (as well as other chemotactic factors), where preincubation with IL-8 reduced the transmigration response to an IL-8 gradient (Figure 4-3).[56] *In vivo,* intravenous IL-8 has been shown to reduce neutrophil accumulation at inflammatory sites[59] (as has GM-CSF[60]), and this probably involves the mechanisms discussed above.

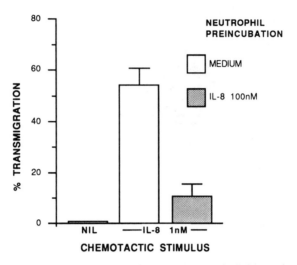

Figure 4-3. Transmigration of neutrophils through endothelial monolayers cultured on filters was stimulated by a gradient of IL-8. Neutrophils preincubated in high doses of IL-8 were inhibited from responding to this gradient. IL-8 preincubated neutrophils responded normally to gradients of fMLP (not shown).

In terms of inflammation, transmigration (and not adhesion) is the most significant feature of neutrophil–endothelial interactions, as it is this step that determines neutrophil tissue infiltration. By this criterion, endogenously-secreted IL-8 is a proinflammatory cytokine. It has been shown that an exogenous gradient of IL-8 promotes transendothelial migration, and even if IL-8 is present on both sides in equal concentrations, there is a modest increase in migration by chemokinetic mechanisms.[55] In fact, a gradient of IL-8 favoring migration into the tissues is likely *in vivo*, given the secretion of IL-8 by activated lymphocytes, monocytes, fibroblasts, and many other cell types[51] at the inflammatory focus. IL-8 injected into skin sites leads to an inflammatory lesion,[61,62] and increased tissue levels of IL-8 have been found in psoriasis, rheumatoid arthritis, and a variety of other diseases.[51,63]

Furthermore, endothelium activated by TNF or IL-1, which has been shown to produce IL-8,[64] promotes neutrophil transmigration,[47,55,65,66] despite inducing loss of neutrophil surface LECAM-1.[42] Neutralizing antibody experiments have provided evidence that the IL-8 produced plays a positive role in this transmigration.[56] It is likely that endothelial IL-8 forms a gradient favoring migration into the tissues, as IL-8 secreted lumenally will be diluted by the serum, whereas it will remain and be concentrated in the subendothelial connective tissue fluid. It is also possible that IL-8 is secreted preferentially in the basal direction (see Chapter 5). Finally, it should be noted that IL-8 has an activating effect on neutrophils and leads to degranulation and reactive oxygen metabolic production,[51] which is a feature of inflammation in tissues.

Therefore, it seems likely that the physiologic role of IL-8 is as a proinflammatory molecule, despite the apparent anti-inflammatory effects of exogenously added IL-8 on assays of adhesion under conditions of flow[42] and assays of transmigration.[56] Interpretation of the *in vitro* and *in vivo* studies depends on whether the IL-8 is on the vascular or tissue side of the endothelium. *In vivo,* tissue IL-8 is more likely to be functionally relevant, and, therefore, IL-8 is clearly a proinflammatory molecule. Pharmacologically-administered intravenous IL-8 may have transient anti-inflammatory effects, but at the expense of activating neutrophils intravascularly, so it is unlikely to achieve therapeutic use for this purpose.

Role of GMP-140 in Cell Adhesion. The GMP-140 present in the α granules of platelets and in the Weibel-Palade bodies of EC[67] is the third member of the LECCAM family of adhesion molecules.[68] Activation of EC with agents such as thrombin and histamine results in translocation of GMP-140 to the cell surface and leads to enhanced neutrophil adhesion, which is mediated, at least in part, through GMP-140.[69,70] Moreover, experiments with purified GMP-140 in lipid bilayers[3] suggest that under flow conditions, GMP-140 may be responsible for the rolling and subsequent arresting of neutrophils seen in the postcapillary venules during an inflammatory response. This binding of neutrophils to GMP-140 is an essential initial event prior to the involvement of other adhesion systems such as ICAM-1 and LFA-1.

The cloning of GMP-140 suggested the possibility that a secreted product may exist, because an mRNA species lacking the transmembrane domain was found.[68] Thus, the possibility that GMP-140 has a cytokine-like action was raised. In order to investigate the function of this putative secreted product, purified GMP-140 from platelets was maintained in solution and added to adhesion assays involving neutrophil–EC interactions. Fluid phase GMP-140 significantly decreased the adhesion of TNF-activated neutrophils to EC (and to plastic surfaces, a process that is ultimately CD18-dependent) but had no effect on the adhesion of "resting" neutrophils to activated EC (mainly ELAM-1-dependent).

Furthermore, fluid phase GMP-140 or GMP-140 immobilized onto solid matrix delays or prevents the release of superoxide anions from activated neutrophils[71] and eosinophils (Vadas, Lucas, Gamble, Skinner, and Berndt, unpublished data). It would be suggested that GMP-140 selectively influences activation processes, all of which apparently involve CD18, resulting in inhibition of adhesion and suppression of the respiratory burst of neutrophils and eosinophils.

GMP-140 has been detected in the plasma of normal individuals and when purified had similar properties to GMP-140 from platelets (Dunlop, Skinner, Bendall, Favalaro, Castaldi, Gorman, Gamble, Vadas, and Berndt, unpublished data). Thus, GMP-140 in the circulation may be an anti-inflammatory molecule in limiting neutrophil and eosinophil activation. However, under appropriate conditions (such as EC activation), this antiinflammatory molecule takes on a proinflammatory role favoring neutrophil arrest[3] and accumulation at inflammatory sites.

Eosinophils

The lineage-specific cytokine IL-5 stimulates eosinophils to adhere to EC by CD11/CD18 molecules.[72] This action appears similar to that of TNF or GM-CSF for neutrophils and is mediated by a conformational change in CD11/CD18. Because IL-5 has no action on myeloid lineages other than eosinophil, it provides a mechanism of selective activation of eosinophils. Platelet-activating factor (PAF), a more general stimulus, has a similar effect on eosinophil adhesion.[73] The ligand for CD11/CD18 on eosinophils is presumed to be ICAM-1 on EC.

Activated EC. It is clear that eosinophils adhere better to IL-1- or TNF-activated EC than to resting EC.[6] The mechanism of this increase involves several adhesion systems. In one report,[24] antibodies to ELAM-1 and ICAM-1 inhibited adhesion. In another,[74] antibodies to ELAM-1, as well as to VCAM, inhibited adhesion 4 hours after activation of the endothelium. In a third,[75] antibodies to ELAM did not diminish eosinophil adhesion, while anti-VCAM antibodies were very effective. And finally, a fourth paper[76] describes a definite role for ELAM-1 and VCAM-1. It can be assumed that positive experiments have more weight than negative ones and that ELAM and VCAM have a definite role in eosinophil adhesion and that the role of ICAM is also likely. Experiments with ICAM have to be treated cautiously in view of the different adhesion specificity of the different Ig domains.[13]

It is important to note that eosinophil infiltration is observed in the LAD syndrome and that IL-4 transgenic mice have striking tissue eosinophilia.[77,78] Taken together, these findings point to a vital role for VCAM–VLA-4 axis in governing eosinophil adhesion and infiltration. The observation that IL-4 increases VCAM but decreases ELAM and ICAM-1[17,31] further provides a mechanism for the preferential accumulation of eosinophils over neutrophils in allergic and helminthic diseases.

Monocytes

Unstimulated monocytes adhere *in vitro* strongly to unstimulated EC, an adhesion that is mediated in part by the CD18 complex.[79] The binding of monocytes to TNF, IL-1, or LPS-stimulated endothelium is increased,[80] and this increase is at least partially due to binding through ELAM-1 and VCAM-1.[81] The difficulty in demonstrating inhibition of adhesion of monocytes to activated endothelium by a single monoclonal antibody[81] and the sometimes contrasting results in different publications may be due to the strong transmigratory stimulus provided by activated endothelium and the necessity to block integrin-mediated events that may play an essential role in this process.

As with other leukocytes, monocyte adherence is further complicated by direct effects of cytokines on monocytes. In this instance, GM-CSF and IL-3,[32,82] rather than TNF and IL-1, are the two powerful stimulators and IL-4 the inhibitor[32] of this process. Interestingly, these agents failed to cause a clear-cut change in the expression of known expression ligands on monocytes,[83] suggesting that the conformational phenotype, or other yet undiscovered molecules, play a part.

SUMMARY

The regulation of adhesion by cytokines is complicated by the imperfections of *in vitro* assays that do not distinguish adhesion from transmigration. Even with this imperfection, profound influences are seen that have been shown to have *in vivo* counterparts. The strong impression is emerging that it will be a combination of cytokines that will determine changes in EC, and the next phase of studies will concentrate on defining the important combinations and their physiologic relevance. Figure 4-4 attempts to summarize the state of current knowledge regarding the role of endogenous mediators in the regulation of leukocyte adhesion.

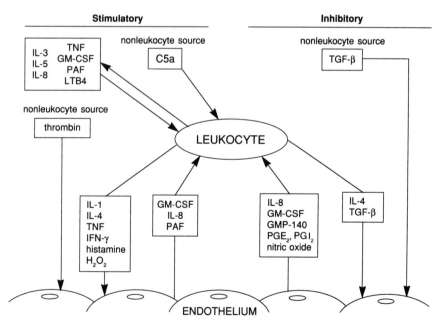

Figure 4-4. Endogenous regulators of leukocyte-endothelial adhesion.

REFERENCES

1. Dustin, M.L., Rothlein, R., Bhan, A.K., Dinarello, C.A., and Springer, T.A. (1986) *J. Immunol.* 137, 245–254.
2. Carlos, T.M., Schwartz, B.R., Kovach, N.L. Yee, E., Rosso, M., Osborn, L., Chi-Rosso, G., Newman, B., Lobb, R., and Harlan, J.M. (1990) *Blood* 76, 965–970.
3. Lawrence, M.B. and Springer, T.A. (1991) *Cell* 65, 859–873.
4. Gamble, J.R., Harlan, J.M., Klebanoff, S.J., and Vadas, M.A. (1985) *Proc. Nat. Acad. Sci. USA* 82, 8667–8671.
5. Bevilacqua, M.P., Pober, J.S., Wheeler, M.E., Cotran, R.S., and Gimbrone, M.A. Jr. (1985) *J. Clin. Invest.* 76, 2003–2011.
6. Lamas, A.M., Mulroney, C.M., and Schleimer, R.P. (1988) *J. Immunol.* 140, 1500–1505.
7. Haskard, D., Cavender, D., Beatty, P., Springer, T., and Ziff, M. (1986) *J. Immunol.* 137, 2901–2906.
8. Carlos, T., Kovach, N., Schwartz, B., Rosa, M., Newman, B., Wayner, E., Benjamin, C., Osborn, L., Lobb, R., and Harlan, J. (1991) *Blood* 77, 2266–2271.
9. Rice, G.E. and Bevilacqua, M.P. (1989) *Science* 246, 1303–1306.
10. Desch, C.E., Dobrina, A., Aggarwal, B.B., and Harlan, J.M. (1990) *Blood* 75, 2030–2034.
11. Staunton, D.E., Marlin, S.D., Stratowa, C., Dustin, M.L., and Springer, T.A. (1988) *Cell* 52, 925–933.
12. Diamond, M.S., Staunton, D.E., Marlin, S.D., and Springer, T.A. (1991) *Cell* 65, 961–971.
13. Diamond, M.S., Staunton, D.E., de Fougerolles, A.R., Stocker, S.A., Garcia-Aquilar, J., Hibbs, M.L., and Springer, T.A. (1990) *J. Cell Biol.* 111, 3129–3139.
14. Osborn, L., Hession, C., Tizard, R., Vassallo, C., Luhowskyj, S., Chi-Rosso, G., and Lobb, R. (1989) *Cell* 59, 1203–1211.
15. Carlos, T.M., Schwartz, B.R., Kovach, N.L., Yee, E., Rosso, M., Osborn, L., Chi-Rosso, G., Newman, B., Lobb, R., and Harlan, J.M. (1990) *Blood* 76, 965–970.
16. Masinovsky, B., Urdal, D., and Gallatin, W.M.(1990) *J. Immunol.* 145, 2886–2895.
17. Thornhill, M.H. and Haskard, D.O. (1990) *J. Immunol.* 145, 865–872.
18. Elias, M.J., Osborn, L., Takada, Y., Elices, M.J., Osborn, L., Takada, Y., Crouse, C., Luhowskyj, S., Hemler, M.E., and Lobb, R. (1990) *Cell* 60, 577–584.
19. Bevilacqua, M.P. and Gimbrone, M.A. Jr.(1989) In Leukocyte Adhesion Molecules. (Springer, T.A., Anderson, D.A., Rosenthal, A.S., and Rothlein, R. eds. Springer Verlag, New York, New York.
20. Bevilacqua, M.P., Pober, J.S., Mendrick, D.L., Cotran, R.S., and Gimbrone, M.A. Jr. (1987) *Proc. Natl. Acad. Sci. USA* 84, 9238–9242.
21. Gamble, J.R., Khew-Goodal, Y.S., and Vadas, M.A. Submitted.
22. Picker, L.J., Kishimoto, T.K., Smith, C.W., Warnock, R.A., and Butcher, E.C. (1991) *Nature* 349, 796–798.
23. Shimizu, Y., Shaw, S., Graber, N., Gopal, T.V., Horgan, K.J., Van Seventer, G.A., and Newman, W. (1991) *Nature* 349, 799–802.

24. Kyan-Aung, U., Haskard, D.O., Poston, R.N., Thornhill, M.H., and Lee, T.H. (1991) *J. Immunol.* 146, 521–528.
25. Phillips, M.L., Nudelman, E., Gaeta, F.C., Perez, M., Singhal, A.K., Hakomori, S., and Paulson, J.C. (1990) *Science* 250, 1130–1132.
26. Tiemeyer, M., Swiedler, S.J., Ishihara, M., Moreland, M., Schweingruber, H., Hirtzer, P., and Brandley, B.K. (1991) *Proc. Natl. Acad. Sci. USA* 88, 1138–1142.
27. Berg, E.L., Robinson, M.K., Mansson, O., Butcher, E.C., and Magnani, J.L. (1991) *J. Biol. Chem.* 266, 14869–14872.
28. Kishimoto, T.K., Warnock, R.A., Jutila, M.A., Butcher, E.C., Lane, C., Anderson, D.C., and Smith, C.W. (1991) *Blood* 78, 805–811.
29. Gamble, J.R. and Vadas, M.A. (1988) *Science* 242, 97–99.
29a. Casscells, W., Bazoberry, F., Speir, E., Thompson, N., Flanders, K., Kondaiah, P., Ferrans, V.J., Epstein, S.E., and Sporn, M. (1990) *Ann. N.Y. Acad. Sci.* 593, 148–160.
30. Thornhill, M.H., Kyan-Aung, U., and Haskard, D.O. (1990) *J. Immunol.* 144, 3060–3065.
31. Thornhill, M.H., Wellicome, S.M., Mahiouz, D.L., Landsbury, J.S., Kyan-Aung, U., and Haskard, D.O. (1991) *J. Immunol.* 146, 592–598.
32. Elliott, M.J., Gamble, J.R., Park, L.S., Vadas, M.A., and Lopez, A.F. (1991) *Blood* 77, 2739–2745.
33. Vedder, N.B. and Harlan, J.M. (1988) *J. Clin. Invest.* 81, 676–682.
34. Buyon, J.P., Abramson, S.B., Philips, M.R., Slade, S.G., Ross, G.D., Weissman, G., and Winchester, R.J. (1988) *J. Immunol.* 140, 3156–3160.
35. Lo, S.K., Detmers, P.A., Levin, S.M., and Wright, S.D. (1989) *J. Exp. Med.* 169, 1779–1793.
36. Nathan, C. and Sanchez, E. (1990) *J. Cell. Biol.* 111, 2171–2181.
37. Gamble, J.R., Rand, T.H., Lopez, A.F., Clark-Lewis, I., and Vadas, M.A. (1990) *Exp. Hematol.* 18, 897–902.
38. Hallmann, R., Jutila, M.A., Smith, C.W., Anderson, D.C., Kishimoto, T.K., and Butcher, E.C. (1991) *Biochem. Biophys. Res. Commun.* 174, 236–243.
39. Spertini, O., Kansas, G.S., Munro, J.M., Griffin, J.D., and Tedder, T.F. (1991) *Nature* 349, 691–694.
40. Griffin, J.D., Spertini, O., Ernst, T.J., Belvin, M.P., Levine, H.B., Kanakura, Y., and Tedder, T.F. (1990) *J. Immunol.* 145, 576–584.
41. Berg, M. and James, S.P. (1990) *Blood* 76, 2381–2388.
42. Smith, C.W., Kishimoto, T.K., Abbass, O., Hughes, B., Rothlein, R., McIntire, L.V., Butcher, E., and Anderson, D.C. (1991) *J. Clin. Invest.* 87, 609–618.
43. Lawrence, M.B., Smith, C.W., Eskin, S.G., and McIntire, L.V. (1990) *Blood* 75, 227–237.
44. Okada, Y., Kawagishi, M., and Kusaka, M. (1990) *Experientia* 46, 1050–1053.
45. Gamble, J.R., Smith, W.B., and Vadas, M.A. in *Tumor Necrosis Factors* (Beutler, B., ed.) Raven Press. In press.
46. Atkinson, Y.H., Marasco, W.A., Lopez, A.F., and Vadas, M.A. (1988) *J. Clin. Invest.* 81, 759–765.
47. Smith, C.W., Marlin, S.D., Rothlein, R., Toman, C., and Anderson, D.C. (1989) *J. Clin. Invest.* 83, 2008–2017.
48. Salyer, J.L., Bohnsack, J.F., Knape, W.A., Shigeoka, A.O., Ashwood, E.R., and Hill, H.R. (1990) *Am. J. Pathol.* 136, 831–841.

49. Mileski, W.J., Winn, R.K., Vedder, N.B., Pohlman, T.H., Harlan, J.M., and Rice, C.L. (1990) *Surgery* 108, 206–212.
50. Horgan, M.J., Wright, S.D., and Malik, A.B. (1990) *Am. J. Physiol.* 259, 1315–1319.
51. Baggiolini, M., Walz, A., and Kunkel, S.L. (1989) *J. Clin. Invest.* 84, 1045–1049.
52. Gimbrone, M.A., Obin, M.S., Brock, A.F., Luis, E.A., Hass, P.E., Hebert, C.A., Yip, Y.K., Leung, D.W., Lowe, D.G., Kohr, W.J., Darbonne, W.C., Bechtol, K.B., and Baker, J.B. *Science* 246, 1601–1603.
53. Wheeler, M.E., Luscinskas, F.W., Bevilacqua, M., and Gimbrone, M.A. (1988) *J. Clin. Invest.* 81, 1211.
54. Hebert, C.A., Luscinskas, F.W., Kiely, J.M., Luis, E.A., Darbonne, W.C., Bennett, G.L., Liu, C.C., Obin, M.S., Gimbrone, M.A., Jr., and Baker, J.B. (1990) *J. Immunol.* 145, 3033–3040.
55. Smith, W.B., Gamble, J.R., Clark-Lewis, I., and Vadas, M.A. (1991) *Immunology* 72, 65–72.
56. Smith, W.B., Gamble, J.R., and Vadas, M.A. Unpublished data.
57. Detmers, P.A., Lo, S.K., Olsen-Egbert, E., Walz, A., Baggiolini, M., and Cohn, Z.A. (1990) *J. Exp. Med.* 171, 1155–1162.
58. Walz, G., Aruffo, A., Kolanus, W., Bevilacqua, M., and Seed, B. (1990) *Science* 250, 1132–1135.
59. Hechtman, D.H., Cybulsky, M.I., Fuchs, H.J., Baker, J.B., and Gimbrone, M.A., Jr. (1991) *J. Immunol.* 147, 883–892.
60. Peters, W.P., Stuart, A.N., Affronti, M.L., Kim, C.S., and Coleman, R.E. (1988) *Blood* 72, 1310–1315.
61. Colditz, I., Zwahlen, R., Dewald, B., and Baggiolini, M. (1989) *Am. J. Path.* 134, 755–760.
62. Rampart, M., Van Damme, J., Zonnekeyn, L., and Herman, A.G. (1989) *Am. J. Path.* 135, 21–25.
63. Seitz, M., Dewald, B., Gerber, N., and Baggiolini, M. (1991) *J. Clin. Invest.* 87, 463–469.
64. Strieter, R.M., Kunkel, S.L., Showell, J.H., Remick, D.G., Phan, S.H., Ward, P.A., and Marks, R.M. (1989) *Science* 243, 1467–1469.
65. Moser, R., Schleiffenbaum, B., Groscurth, P., and Fehr, J. (1989) *J. Clin. Invest.* 83, 444–455.
66. Furie, M.B. and McHugh, D.D. (1989) *J. Immunol.* 143, 3309–3317.
67. McEver, R.P., Beckstead, J.H., Moore, K.L., Marshall-Carlson, L., and Bainton, D.F. (1989) *J. Clin. Invest.* 84, 92–99.
68. Johnston, G.I., Cook, R.G., and McEver, R.P. (1989) *Cell* 56, 1033–1044.
69. Geng, J.G., Bevilacqua, M.P., Moore, K.L., McIntyre, T.M., Prescott, S.M., Kim, J.M., Bliss, G.A., Zimmerman, G.A., and McEver, R.P. (1990) *Nature* 343, 757–760.
70. Toothill, V.J., Van Mourik, J.A., Niewenhuis, H.K., Metzelaar, M.J., and Pearson, J.D. (1990) *J. Immunol.* 145, 283–291.
71. Wong, C.S., Gamble, J.R., Skinner, M.P., Lucas, C.M., Berndt, M.C., and Vadas, M.A. (1991) *Proc. Natl. Acad. Sci. USA* 88, 2397–2401.
72. Walsh, G.M., Hartnell, A., Wardlaw, A.J., Kurihara, K., Sanderson, C.J., and Kay, A.B. (1991) *Immunol.* 71, 258–265.

73. Kimani, G., Tonnesen, M.G., and Henson, P.M. (1988) *J. Immunol.* 140, 3161–3166.
74. Bochner, B.S., Luscinskas, F.W., Gimbrone, M.A., Jr., Newman, W., Sterbinsky, S.A., Derse-Anthony, C.P., Klunk, D., and Schleimer R.P. *J. Exp. Med.* 173, 1553–1556.
75. Dobrina, A., Menegazzi, R., Carlos, T.M., Nardon, E., Cramer, R., Zacchi, T., Harlan, J.M., and Patriarca, P. (1991) *J. Clin. Invest.* 88, 20–26.
76. Weller, P.F., Rand, T.H., Goelz, S.E., Chi-Rosso, G., and Lobb, R.R. (1991) *Proc. Natl. Acad. Sci. USA* 88, 7430–7433.
77. Tepper, R.I., Pattengale, P.K., and Leder, P. (1989) *Cell* 57, 503–512.
78. Tepper, R.I., Levinson, D.A., Stanger, B.Z., Campos-Torres, J., Abbas, A.K., and Leder, P. (1990) *Cell* 62, 457–467.
79. Carlos, T.M. and Harlan, J.M. (1990) *Immunol. Rev.* 114, 5.
80. Carlos, T.M., Dobrina, A., Ross, R., and Harlan, J.M. (1990) *J. Leuk. Biol.* 48, 451–456.
81. Carlos, T., Kovach, B., Schwartz, B., Rosa, M., Newman, B., Wagner, E., Banjamin, C., Osborn, L., Lobb, R., and Harlan, J. (1991) *Blood* 77, 2266–2271.
82. Gamble, J.R., Elliott, M.J., Jaipargas, E., Lopez, A.F., and Vadas, M.A. (1989) *Proc. Nat. Acad. Sci. USA* 86, 7169–7173.
83. Elliott, M.J., Lopez, A.F., Gamble, J.R., and Vadas, M.A. Unpublished observations.

Acknowledgments: Supported by grants from the National Health and Medical Research Council of Australia.

CHAPTER 5

Transendothelial Migration

C. Wayne Smith

A critical step in the inflammatory process is migration of circulating leukocytes from the vascular compartment through the endothelial junctions and perivascular structures into tissues. This phenomenon has been documented morphologically in a variety of models *in vivo*.[1-5] The discussion that follows deals with the possible contributions of specific adhesion molecules to this process and will focus primarily on the mechanisms used by neutrophils, as this cell type has been more extensively studied than other leukocytes.

TRANSENDOTHELIAL MIGRATION *IN VITRO*

The migration of neutrophils through confluent monolayers of endothelial cells *in vitro* has been investigated as a possible model for the phenomenon of leukocyte emigration at sites of acute inflammation. Beesley et al.[6] examined the interaction of porcine neutrophils with visually confluent porcine endothelial monolayers on glass coverslips.[7] They observed that a portion of the neutrophils contacting the apical endothelial cell surface migrated between the endothelial cell junctions to a position beneath the monolayer. These cells could be easily distinguished from the nonmigrating cells under phase contrast microscopy by their markedly flattened appearance in the focal plane of the basal surface of the monolayer (Figure 5-1). The migration was rapid, with individual cells proceeding from above to below the monolayer within 30 seconds. Electron microscopic studies revealed a sequence of events very similar to those seen with migrating neutrophils *in vivo*. The initial contact between neutrophils and endothelium was apparently mediated by microvillus-like extensions of organelle-free neutrophil cytoplasm, while the main body of the cell remained roughly spherical and out of contact. Subsequent granulocyte contact involved the close apposition of areas of the blood cell to the endothelial surface, creating small indentations, and at points of closest contact bundles of filaments were often found in the endothelial cytoplasm just beneath the plasma membrane. Neutrophils located above the area where two endothelial cells abutted were often seen with pseudopodia extruded, separating the two endothelial cells. Examination of leukocytes in various stages of migration led the authors to conclude that the cytoplasm including granules flowed beneath the monolayer first, and the portion of the cell containing the nucleus followed last. The flattening of the granulocyte beneath the endothelium was maintained after migration was complete.

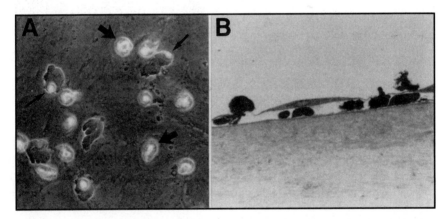

Figure 5-1. Transendothelial migration of neutrophils. Visually confluent monolayers of human umbilical vein endothelial cells were grown on fibronectin-coated glass coverslips and stimulated with recombinant IL-1β for three hours. Isolated human neutrophils were allowed to settle onto the coverslips for approximately five minutes at 37°C. (A) When viewed under phase contrast microscopy, neutrophils in the focal plane near the glass surface were flattened and bipolar in shape, some with their uropods protruding through to the apical surface of the monolayer (small arrows). Neutrophils on the apical surface of the monolayer were out of focus (large arrow). (B) As seen in cross section, the neutrophils were both above and beneath the monolayer, and one cell is seen with its uropod still protruding through to the apical surface. (Smith et al., (1988) *J. Clin. Invest.* 82, 1746)

While such observations supported earlier work[8] showing that neutrophil migration *in vitro* is not inhibited by intercellular contact and that neutrophils adhered sufficiently well to endothelial cells to crawl across their surface, it provided little insight into the mechanisms that bring about this migration. Indeed, leukocyte locomotion in these studies appeared to be spontaneous. Taylor et al.[9] modified the culture conditions to allow for a chemotactic gradient across the endothelial monolayer. They established confluent bovine aortic endothelial cell monolayers on gelatin-treated polycarbonate filters, inserted them into chemotaxis chambers, and demonstrated that ^{51}Cr-labelled human neutrophils would migrate into the lower compartment of the chamber containing the chemotactic tripeptide, f-Met-Leu-Phe (10 nM). Migration could be detected within 30 minutes and continued throughout the 90-minute observation period. Very few neutrophils were found in the lower compartment if f-Met-Leu-Phe was omitted, suggesting that under these culture conditions transendothelial migration was not a spontaneous event.

Because these techniques involved the culture of endothelial cells on artificial surfaces, there was concern that some important features of the vessel wall that might physically impede or otherwise alter the migration of neutrophils were not reproduced in these models. Furie et al.[10] prepared confluent monolayers of bovine microvascular endothelial cells on human amniotic membrane. These monolayers restricted the passage of macromolecules and electrical current. When a concentration gradient of f-Met-Leu-Phe or leukotriene B_4 (LTB_4) was established across these membranes, transendothelial migration of isolated human neutrophils was markedly increased.[11] Because the evaluation of neutrophil migration was made visually on cross-sections of monolayer preparations, it was possible to actually determine that migration was very rare in the absence of chemotactic factors. These investigators found that in response to f-Met-Leu-Phe (10 nM), neutrophils adhered to the endothelium within two minutes and began to migrate within five minutes. At 10 minutes, the majority of the migrating cells were beneath the endothelium. In this model, between 20% and 40% of the neutrophils added to the chamber eventually migrated across the endothelial monolayer, and those neutrophils that did not migrate either failed to adhere or detached from the apical surface of the endothelial cells. Transmission electron microscopic studies revealed some of the same features seen in the studies of Beesley et al.[6] Pseudopodia of the adhering neutrophils appeared to indent the apical plasma membranes of the endothelial cells, but migration began by insertion of a pseudopod between two endothelial cells with the neutrophils maintaining close contact with the adjacent endothelial cells as they traversed the monolayer (Figure 5-2). In contrast to the previous studies, endothelial cells grown on the stromal surface of the amniotic membrane produced an underlying basement membrane, and following emigration, the neutrophils were observed lying between the endothelium and this membrane tightly apposed to the basal surface of the endothelial cells, a phenomenon repeatedly seen in studies *in vivo*.[2,3,5,12,13] With time, the neutrophils moved past the basement membrane into the amniotic tissue.

Another variation was developed by Huber and Weiss.[14] They prepared confluent monolayers of human umbilical vein endothelial cells (HUVECs) on type I collagen gels on polycarbonate filters and demonstrated that after 21 days in culture these monolayers produced a continuous basement membrane with biochemical and structural features of subendothelial basement membranes *in vivo*. As observed in the Furie et al. study,[11] transmigration of neutrophils did not occur unless a chemotactic gradient was established across the monolayer, and then, approximately 40% of the neutrophils contacting the apical surface of the monolayer migrated past the endothelial cells into the collagen gel. Also, neutrophil migration appeared to be transiently impeded by the basement membrane.

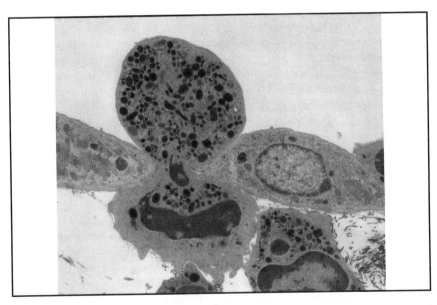

Figure 5-2. Transendothelial migration of neutrophils. Confluent monolayers of bovine microvascular endothelial cells were grown on amnionic membrane and mounted in a chamber with compartments above and below the membrane. The chemotactic peptide, f-Met-Leu-Phe (100 nM) was placed below the membrane and isolated human neutrophils were placed above. This transmission electron micrograph demonstrates a neutrophil in the process of transmigration. (Furie et al. [1987] *J. Cell Sci.* 88, 161).

TRANSMIGRATION INDUCED BY CYTOKINE STIMULATION OF THE ENDOTHELIUM

Numerous studies *in vivo* indicate that interleukin-1 (IL-1) and tumor necrosis factor-α (TNF-α) may play an important role in the extravasation of neutrophils.[15–22] While these studies deal with the accumulation of neutrophils at the site of inflammation, they do not distinguish effects on leukocyte adhesion to the luminal surface of the vascular endothelium from effects on subsequent leukocyte migration. Accumulation of leukocytes at a site of inflammation could be due to either function, and these cytokines are well known to increase the adhesiveness of cultured endothelial cells for previously unstimulated neutrophils.[23] This question has been investigated using variations of the *in vitro* models described above. Smith et al.[24] used a variation of the Beesley et al. model.[6] They prepared visually confluent monolayers of HUVECs on fibronectin-coated glass coverslips and allowed isolated human neutrophils to settle onto these monolayers. Direct visualization using phase contrast microscopy revealed that in contrast to the observations of Beesley et al., the isolated neutrophils were spherical and remained spherical after contacting monolayers of unstimulated HUVECs over a 30-minute observation period. However, if the HUVEC monolayers had been exposed previously to IL-1 (0.3–3.0 U/ml) or endotoxin lipopolysaccharide (LPS) (1–10 ng/ml) for three hours, a high percentage of the contacting neutrophils would rapidly change shape and migrate beneath the monolayer, there assuming the flattened morphology observed by several other investigators.[6,25–28] Under basal conditions, less than 1% of the neutrophils exhibited this behavior, but between 35% and 78% of the cells migrated following cytokine stimulation of the endothelial cells.

Moser et al.[29] used a modification of the technique of Taylor et al.,[9] preparing HUVEC monolayers on polycarbonate filters mounted in a chemotaxis chamber, and determined transendothelial migration by counting the number of neutrophils within the lower compartment at the end of incubation. Between 20% and 40% of the neutrophils added to the upper compartment migrated through the junctions of the endothelial cell monolayer if the monolayer had been preincubated with IL-1 (1 U/ml) or TNF (5 ng/ml). This effect of IL-1 was inhibited by actinomycin D. Scanning electron microscopy revealed a change to a polarized motile shape of neutrophils contacting cytokine-stimulated monolayers, and evidence was provided that transendothelial migration was inhibited by pertussis toxin pretreatment of the neutrophils. These and other observations led the investigators to conclude that the stimulated endothelial cells provoke the neutrophil transmigration.

Furie and McHugh[30] evaluated neutrophil migration using confluent HUVEC monolayers grown on human amniotic membrane. They confirmed the observations of Smith et al.[24] and Moser et al.[29] that stimulation of endothelial cells with IL-1 and TNF-α resulted in increased transendothelial migration. The time course of this migration was similar to that induced by a chemotactic gradient, being almost complete by 10 minutes. Migration was very low in the absence of stimulation (approximately 3% of the added neutrophils migrated), and as observed in most of the published studies *in vitro*, between 20% and 40% of the added neutrophils migrated across activated endothelial cell monolayers. The reason for this partial response is not clear. In most studies, the nonmigrating cells either did not attach to the apical surface or transiently adhered. In fact, this population could not be recruited to migrate if a chemotactic gradient was established across cytokine-stimulated monolayers.[29,30] Furie et al.[11,30] provided evidence that the endothelial monolayer is not limiting penetration under these experimental conditions, as the number of neutrophils migrating per unit area of endothelium was directly proportional to the number added to the chamber over a wide range for both chemotactic and endothelial-induced migration.

Most recently, Luscinskas et al.[31] have demonstrated that IL-1 and TNF-α stimulation of HUVEC monolayers induces transendothelial migration. They used an experimental model similar to that of Huber and Weiss,[14] growing the monolayer on type I collagen gel on a clear plastic surface for direct visualization of the migratory process. At maximum migration, approximately 50% of the neutrophils contacting the cytokine-stimulated endothelium penetrated the monolayer, with most of the migration occurring between 10 and 20 minutes after contact. Maximum migration occurred at four hours after IL-1 stimulation of the endothelial cells, and little transmigration was seen at 24 and 48 hours. Direct observation revealed that most neutrophils on the apical surface of unstimulated monolayers retained a rounded appearance throughout a 20-minute observation period, but those on stimulated endothelium assumed shapes consistent with neutrophil locomotion within one to three minutes.

CYTOKINE-STIMULATED ENDOTHELIAL CELLS ACTIVATE NEUTROPHIL MOTILITY

The apparent activation of neutrophils following contact with cytokine-stimulated neutrophils is unlikely to be a direct effect of IL-1 or TNF-α, because in most experimental protocols the endothelial cells were washed prior to introduction of the neutrophils, and current evidence indicates that IL-1 is incapable of directly stimulating neutrophil locomotion.[32,33] One likely possibility is that these cytokines induce endothelial cells to express chemotactic substances. Several studies show that endothelial cells are capable of producing such substances following a variety of stimuli.[34-45] Two factors are of particular interest, because they are produced by cytokine-stimulated endothelial cells within the time during which neutrophil transmigration is provoked by the monolayers. Platelet activating factor (PAF) may be produced by HUVEC following IL-1 and TNF-α stimulation,[35] though this is not altogether clear.[46] PAF is known to be chemotactic for neutrophils[47] and has recently been implicated in the extravasation of neutrophils in mesenteric vessels following ischemia and reperfusion.[48] IL-8 is produced by HUVECs for an extended time following cytokine stimulation[38] and is well known as a chemotactic factor for neutrophils.[49–52] Huber et al.[52a] have recently used the *in vitro* model developed by Huber and Weiss[14] in evaluating the contribution of IL-8 to the transmigration of neutrophils induced by IL-1β, TNF-α or LPS stimulation of human endothelial cells. They found a correlation between the ability of the endothelial monolayers to synthesize IL-8 and their ability to promote neutrophil transmigration, and immunolocalization of IL-8 revealed that it was in association with both the endothelial cells and the underlying structures supporting the monolayer (basement membrane and type 1 collagen gel). They found that antiserum to IL-8 inhibited neutrophil transmigration by greater than 63%.

There is no direct evidence to indicate that the stimulated endothelial cells create a chemotactic gradient across the monolayer toward which the neutrophils migrate. Moser et al.[29] found that when C5a was added with the neutrophils above the endothelial monolayer (i.e., in a position to create a negative gradient), IL-1-stimulated monolayers were still able to induce significant increases in transmigration. Such results are difficult to interpret and may reflect simply an effect of the chemokinetic influence of the putative endothelial factors. Several investigators have found that once the neutrophils pass through the endothelial junctions, they continue to migrate into the matrix supporting the monolayer, and thus away from the monolayer. This suggests chemokinetic rather than directed migration.[14,29–31]

CONTRIBUTIONS OF ADHERENCE TO TRANSENDOTHELIAL MIGRATION

In most of the studies discussed above, the investigators found that less than 50% of the neutrophils contacting the apical surface of endothelial cell monolayers penetrated the monolayer under maximal stimulus conditions. Furie et al.[11,30] observed that the nonmigrating cells either did not adhere to or eventually detached from the monolayer and that adhesion of the migrating population was evident within a few minutes following their contact with the monolayer. Stimuli such as chemotactic factors not only induce transendothelial migration but lead to increased adhesion of neutrophils to the endothelium. Gimbrone et al.[27] found that LTB_4 at concentrations as low as 0.1 nM increased the attachment of ^{111}In-labelled human neutrophils to monolayers of bovine aortic endothelial cells on glass or plastic coverslips when added for 10 minutes to the cultures. This effect appeared not to be species specific, as LTB$_4$ increased human neutrophil attachment to HUVECs, bovine vena cava, human and baboon saphenous vein, and SV-40-transformed HUVECs. Tonnesen et al.[28] demonstrated that f-Met-Leu-Phe, C5a, and C5a des arg were each effective in increasing human neutrophil adherence to unstimulated HUVEC monolayers grown on plastic when added to the cell cultures at 10 nM. This increase was not specific for endothelium, as neutrophil adhesion to human skin fibroblasts and smooth muscle cells was also increased as well. These investigators' adhesion assay was designed to study low affinity adhesions and used static conditions and immediate glutaraldehyde fixation of adherent cells. They found that the onset of stimulated adhesion occurred within 30 seconds of adding the stimulus, was maximal at two minutes, and persisted for up to 45 minutes. Because preincubated and washed neutrophils showed augmented adhesion while preincubated and washed endothelium did not, they concluded that the chemotactic factors acted primarily on the neutrophils. Others have confirmed this adherence-enhancing effect of chemotactic factors,[53–58] and of particular interest are the observations that PAF[28,59] and IL-8,[50,60] factors that are produced by stimulated endothelial cells, are both active in this regard.

While such studies clearly show the potential of chemotactic factors to augment neutrophil adhesion to endothelial cells, other studies indicate that the role of chemotactic factors may be rather complex. Charo et al.[53] confirmed that C5a could increase neutrophil adherence but noted a reciprocal relationship between chemotactic migration and adherence to endothelial cells. Concentrations of C5a (e.g., 1 nM) that induced optimal chemotaxis (i.e., migration through micropore filters toward a higher concentration of C5a) reduced neutrophil adhesion to endothelial monolayers, while concentrations of C5a that inhibited migration (e.g., 10 nM) significantly increased adherence. Tonnesen et al.[28] found that C5a did not increase adherence at concentrations less than 10 nM. Gimbrone et al.[38,61] and Smith et al.[62–64] have also found that chemotactic factors such as IL-8, C5a, f-Met-Leu-Phe, and PAF significantly reduced neutrophil adherence to endothelial monolayers. Thus it is evident that a simple cause and effect relationship between chemotactic stimulation and neutrophil–endothelial adhesion or transendothelial migration does not exist.

Two additional issues may be pertinent to this discussion. The first is that the proadhesive effects of chemotactic stimulation appear to be transient, peaking in five to ten minutes and waning within 30 minutes.[65] The second is that one of the possible endothelial-derived chemotactic factors, PAF, is largely retained by the endothelial cells[35,45] and presented on the plasma membrane.[45] There it may act as a surface-bound stimulus[45] in the microenvironment of the vessel wall where fluid flow may well preclude the establishment of a stable gradient of soluble chemotactic factors. Zigmond[66,67] has clearly shown that a stable gradient of soluble chemotactic factor is required for directional migration *in vitro.*

SPECIFIC ADHESION MOLECULES AND MIGRATION TOWARD CHEMOTACTIC GRADIENTS

CD18 integrins. A number of case reports from several laboratories have described children with recurrent bacterial infections, marked leukocytosis, marked reductions in the migration of neutrophils into inflamed tissue, and an apparent abnormality of neutrophil adhesion *in vitro.*[68–78] This syndrome is caused by a deficiency of CD18[79,80] which prevents or profoundly reduces the surface expression of functional heterodimers LFA-1 (CD11a/CD18), Mac-1 (CD11b/CD18), and p150,95 (CD11c/CD18).[81] Harlan et al.[76] used the model described by Taylor et al.[9] to investigate the ability of neutrophils from two of these patients to migrate through endothelial monolayers toward a chemotactic gradient. While up to 46% of normal neutrophils transmigrated toward f-Met-Leu-Phe in this study, 9% or less of the patient neutrophils migrated. They also demonstrated that a monoclonal antibody to CD18 (60.3) reduced the transendothelial migration of normal neutrophils to the same low level as that of the CD18-deficient neutrophils and postulated that the absence of the CD18 glycoproteins prevents neutrophil adherence to endothelium following neutrophil activation by chemotactic stimuli, thereby accounting for the failure of these patients' neutrophils to emigrate at sites of inflammation.

CD11b/CD18 (Mac-1, Mo1, CR3). Schmalstieg et al.[82] found that chemotactic migration of CD18-deficient neutrophils on a protein-coated plastic surface *in vitro* was less than 20% of that of normal cells. Anderson et al.[83] and Dana et al.[84] demonstrated that chemotactic migration on artificial surfaces was greatly diminished by monoclonal antibodies against CD18 and that the specific heterodimer of greatest importance in this experimental setting was CD11b/CD18, as monoclonal antibodies against CD11b were markedly inhibitory while those against CD11a were not. In addition, stimuli such as chemotactic factors and phorbol esters increase the adhesive functions of constitutive surface CD11b/CD18,[85,86] and they promote rapid (within minutes) translocation of CD11b/CD18 to the cell surface from intracellular stores in the secondary granules.[87–93] Furthermore, the increased adhesion to endothelial cells *in vitro* following chemotactic stimulation of neutrophils is greatly inhibited by anti-CD11b monoclonal antibodies.[54–56,58,94]

Several investigators have proposed that chemotactic stimulation of neutrophils influences the cell surface distribution of adhesion sites, promoting not only the adherence of the advancing lamellipodium to the substratum but the transport of the adhesion sites to the uropod.[95–99] Francis et al.[95] proposed a model whereby sequential expression of cell surface CD11b/CD18 would contribute to the locomotion of adherent neutrophils. They evaluated the surface distribution of CD11b/CD18 on migrating neutrophils using fluorescein- and rhodamine-labelled anti-CD11b applied sequentially to chemotactically stimulated cells and provided evidence that CD11b/CD18 is upregulated during locomotion as CD11b/CD18-positive granules fuse with the plasma membrane near the lamellipodia. The newly expressed CD11b/CD18 is then swept to the uropod as the cell migrates, and accumulates on the retraction fibers. The observations that CD11b/CD18-dependent adhesion sites exhibit the same pattern of appearance and redistribution[83,97] supports the hypothesis that such a mechanism may contribute to neutrophil locomotion. However, current evidence does not support the view that newly upregulated CD11b/CD18 participates in adhesion.[65,100–105] Rather, it seems that constitutive CD11b/CD18 undergoes a qualitative change following stimulation of the neutrophil, which promotes its participation in adhesion. Dransfield and Hogg[106] have obtained evidence for a conformational change in the heterodimer recognized by a monoclonal antibody that defines a Mg++-dependent and temperature-sensitive epitope, and Detmers et al.[85] have provided evidence that phorbol ester stimulation of adherent neutrophils results in transient clustering of CD11b/CD18, a phenomenon that would likely increase the avidity of CD11b/CD18-dependent binding. Thus, while both qualitative and quantitative changes in surface Mac-1 have been documented following exposure of neutrophils to chemotactic factors, their exact roles in neutrophil locomotion remain unknown.

CD11a/CD18 (LFA-1). The contribution of CD11b/CD18 to neutrophil transendothelial migration in a chemotactic gradient was evaluated by Furie et al.[107] using the model with HUVECs grown on amniotic membrane.[11] They evaluated the effects of anti-CD11b monoclonal antibodies previously shown to markedly inhibit adhesion to and migration of neutrophils across protein-coated surfaces and found that they inhibited transendothelial migration through unstimulated endothelial monolayers by less than 40%. In addition, they found that anti-CD11a monoclonal antibody, previously shown to be ineffective in reducing adhesion to and migration across protein-coated artificial surfaces, inhibited transendothelial migration by greater than 50%. The combination of anti-CD11a and anti-CD11b antibodies was almost as inhibitory as anti-CD18. Thus, it appears that in contrast to migration on artificial surfaces, CD11a/CD18 appears to be sufficient for substantial chemotactic migration of neutrophils through confluent endothelial monolayers, though both CD11a/CD18 and CD11b/CD18 are needed for optimum migration.

ICAM-1 (CD54). The known ligands for CD11a/CD18, intercellular adhesion molecules-1 and -2 (ICAM-1, ICAM-2) are constitutively expressed on unstimulated endothelial cells both in culture and *in vivo*.[108–113] Isolated, unstimulated neutrophils are capable of using CD11a/CD18 to adhere to isolated ICAM-1 densely inserted in a planar artificial membrane, but they adhere very little to resting endothelial cell monolayers unless stimulated (e.g., with phorbol ester or chemotactic factors). In the Furie et al. study,[107] anti-ICAM-1 monoclonal antibody, R6.5, inhibited migration to the same degree as anti-CD11a, and inhibition was no greater when anti-ICAM-1 was combined with anti-CD11a. However, when anti-ICAM-1 was combined with anti-CD11b, the inhibition was apparently additive. Such results suggest that under chemotactic conditions, transmigration across unstimulated endothelial monolayers involves CD11a/CD18-ICAM-1 interactions, but a ligand for CD11b/CD18 remains to be defined. Similar results with adhesion to endothelial cells were obtained following stimulation of neutrophils with phorbol esters (i.e., a portion of the adhesion was shown to depend on CD11a/CD18[65] interacting with constitutive ICAM-1 on the endothelial cell.[94,114]). There is no intracellular storage pool for CD11a/CD18 within the neutrophil, and stimulation does not alter cell surface levels of this heterodimer.[115] Thus, stimulation of the neutrophil appears to induce a qualitative change in the ability of CD11a/CD18 to interact with endothelial ICAM-1, and because the resulting adhesion is transient (i.e., peaking within 15 minutes and returning to baseline within 30 minutes), the qualitative change appears to be reversible.[65] Lymphocyte CD11a/CD18-ICAM-1 interactions can be rapidly modulated without changes in the density of CD11a/CD18 on the cell surface by stimulation with phorbol esters or cross-linking certain structures on the cell surface (e.g., T-cell receptor).[116–123] The time course of this change in avidity is similar in stimulated neutrophils and lymphocytes, peaking within 15 minutes and returning to baseline within 30 minutes,[65,116] and is recognized in both cell types (as is the case for CD11b/CD18 on neutrophils) by the appearance of a Mg++-dependent, temperature-sensitive epitope.[106] Such a modulation in CD11a/CD18 may allow neutrophils to adhere to and detach from endothelial ICAMs in the process of transmigration. Recent evidence indicates that transendothelial migration of lymphocytes *in vitro* involves CD11a/CD18-ICAM-1 interactions as well.[124,125]

These studies indicate that chemotactically-stimulated neutrophils can use both CD11a/CD18 and CD11b/CD18 for locomotion. CD11b/CD18 appears to provide the neutrophil with greater flexibility, allowing adhesion to and migration across artificial surfaces and tissue cells. It binds to an array of different proteins and thereby allows neutrophils to adhere to a wide variety of surfaces.[94,126–133] Though not yet proven, the intracellular pool appears to provide replacement for CD11b/CD18 lost from the cell surface during locomotion.[95] In contrast, CD11a/CD18 plays a more limited role, apparently allowing adhesion to and migration over surfaces bearing ICAM-1, and possibly ICAM-2.

SPECIFIC ADHESION MOLECULES AND MIGRATION INDUCED BY CYTOKINE STIMULATION OF THE ENDOTHELIAL CELLS

As discussed above, emigration of neutrophils *in vivo* is unlikely to be simply the result of a chemotactic gradient across unstimulated endothelium, though intradermal injection of C5a des arg, f-Met-Leu-Phe, and LTB_4 in rabbits results in rapid emigration of neutrophils that is maximum within 30 minutes,[134] a time frame that would not allow for *de novo* synthesis of inflammatory determinants. Intradermal injection of IL-1 or TNF-α produces leukocyte emigration that is delayed by 30 to 60 minutes in onset and peaks within two to three hours.[134] Inhibition of protein synthesis largely blocks cytokine-induced extravasation, though it has no effect on the emigration that follows injection of chemotactic agents.[19,134] Protein synthesis-independent and -dependent mechanisms may contribute to different phases of leukocyte localization (e.g., rapid margination after injury[135] followed by more sustained efflux of leukocytes). While the experimental models *in vitro* that provide stable chemotactic gradients reveal some leukocyte and endothelial functions in isolation, they fail to take into account newly upregulated endothelial adhesion molecules and the effects of flow on the interactions of leukocytes and endothelial cells. This point is clearly illustrated by the demonstration that chemotactic stimulation does not increase adhesion of neutrophils to unstimulated endothelial cells under conditions of flow until the wall shear stress is reduced to values below those seen in postcapillary venules,[136,137] the site where most leukocyte emigration occurs. In contrast, adhesion of neutrophils to thrombin- or cytokine-stimulated endothelial cell monolayers *in vitro* does occur at wall shear stresses within the low venous range.[62,64,136,138] A high percentage of the neutrophils that adhere under flow to cytokine-stimulated monolayers migrate beneath the monolayer.[62,64,136,138] Therefore, chemotactic stimulation of neutrophils alone may be insufficient to localize these cells at an inflammatory site without coincident expression of some rapidly mobilized or newly synthesized endothelial molecules.

CD18 integrins. The transendothelial migration of neutrophils provoked by cytokine-stimulated endothelial monolayers *in vitro*[24,29–31] essentially fails to occur when neutrophils from patients with CD18 deficiency are used,[24,56,64,139] and anti-CD18 monoclonal antibodies almost completely inhibit the transmigration of normal neutrophils.[24,31,56,62,64,136,138] The relative contributions of CD11b/CD18 and CD11a/CD18 are not clear. In experiments using a modification of the *in vitro* model developed by Beesley et al.,[6] anti-CD11b monoclonal antibodies produced less than 40% inhibition of migration, anti-CD11a monoclonal antibodies produced greater than 50% inhibition[56,139] and combining the antibodies gave additive inhibitory effects equivalent to anti-CD18. Using the model developed by Luscinskas et al.,[31] antibodies against either CD11a or CD11b produced greater than 90% inhibition of transmigration. The results using both models indicate that both CD11a/CD18 and CD11b/CD18 are needed for optimal transmigration, a conclusion consistent with the results of Jutila et al.[140] *in vivo*. They found that systemic administration of either anti-CD11a or anti-CD11b monoclonal antibodies significantly inhibited (by greater than 60%) the influx of neutrophils into the thioglycollate-inflamed peritoneal cavity of mice.

CD11c/CD18 (p150,95). Relatively little is known about the contribution of this β_2 integrin to the adhesion of neutrophils to endothelial cells. Stacker and Springer[141] isolated the functional heterodimer and demonstrated that when adsorbed to a plastic surface, CD11c/CD18 would support the adhesion of endothelial cells. Unstimulated endothelial cells bound poorly but stimulation of these cells with IL-1β or endotoxin for 18 hours caused a significant increase in the level of binding to CD11c/CD18. This binding could be specifically inhibited with antibodies to CD11c. The endothelial ligand remains undefined. Because neutrophils and monocytes have this integrin on their surfaces, it is reasonable to conclude that it may participate in the adhesion and migration of these leukocytes through activated endothelium. Current published evidence supports this possibility, though its contribution to adhesion appears relatively small.[142–145]

ICAM-1 (CD54). Among the numerous changes in endothelial cells that occur following cytokine stimulation,[23] the increase in adhesiveness for unstimulated neutrophils seen in the first few hours is largely attributable to synthesis and surface expression of ICAM-1[24] and endothelial leukocyte adhesion molecule-1 (ELAM-1).[146–147] When evaluated at the peak time for neutrophil adhesion following IL-1 or TNF-α stimulation (3–4 hours), anti-ICAM-1 monoclonal antibodies R6.5[24] and Hu5/3[31] each produced partial inhibition (30%–50%), and anti-ELAM-1 monoclonal antibodies CL2 and CL3[63,148] and H18/7[31,146] each produced partial inhibition (30%–50%). Combined use of anti-ICAM-1 and anti-ELAM-1 antibodies causes essentially additive inhibition of adhesion[31,63] (70%–80%). Such results indicate that ICAM-1 and ELAM-1 are components of distinct adherence pathways.

Since transendothelial migration appears to require the β_2 integrins, principally CD11a/CD18 and CD11b/CD18, endothelial ICAM-1 should also play an important role. As discussed above, LFA-1–ICAM-1 interactions apparently account for a portion of the transmigration of neutrophils in a chemotactic gradient,[107] but the ligand for CD11b/CD18 under those conditions has not been defined. Experimental protocols using isolated ICAM-1 densely inserted in planar artificial membranes have shown that isolated neutrophils exposed to a chemotactic factor exhibit a high degree of binding, and that anti-CD11b monoclonal antibodies reduce this adherence by approximately 50%.[56,139] Diamond et al.[149] more directly addressed the question of CD11b/CD18 binding to ICAM-1 by demonstrating that COS cells transfected with both cDNAs for CD18 and CD11b bound to isolated ICAM-1, and that anti-ICAM-1 monoclonal antibody, R6.5,[24] was almost completely inhibitory. The reciprocal experiment was also performed showing that COS cells and HUVEC expressing ICAM-1 bound to planar membranes containing isolated CD11b/CD18 heterodimer. The interaction of CD11b/CD18 and CD11a/CD18 with ICAM-1 was not found to be identical, though. Monoclonal antibodies recognizing different domains of ICAM-1 exhibited differential ability to inhibit these interactions (e.g., R6.5[24] binding to domain 2 inhibited CD11b/CD18 and LB2[150] binding to domain 1 inhibited CD11a/CD18), and the binding of CD11b/CD18 with ICAM-1 was less shear resistant than bound CD11a/CD18. Furthermore, using domain-deleted and chimeric forms of ICAM-1, Staunton et al.[151] demonstrated that CD11a/CD18 bound to the first domain of ICAM-1, and Diamond et al.[149] demonstrated that CD11b/CD18 bound to the third domain of ICAM-1.

While the functional significance of the interaction of CD11b/CD18 with ICAM-1 is not clear, there is evidence suggesting that it contributes to transendothelial migration induced by cytokine stimulation of the endothelial cells.[56] Diamond et al.[149] suggested that CD11b/CD18 and CD11a/CD18 on the same leukocyte may be able to bind to a single ICAM-1 molecule, and they provided information on the structural feasibility of such an interaction. In addition, they found that the size of the N-linked oligosaccharide chain on domain 3 of ICAM-1 affects the interaction between CD11b/CD18 and ICAM-1. Adhesion depends primarily on Mac-1 if ICAM-1 has a smaller N-linked side chain but on CD11a/CD18 if ICAM-1 has a complex carbohydrate side chain. ICAM-1 from different cell types has been shown to differ in the level and type of N-linked carbohydrates,[109] and thereby may regulate biological interactions of different cells with neutrophils. For example, Entman et al. have found that canine neutrophils adhere to isolated, cytokine-stimulated canine cardiac myocytes[59] primarily through a mechanism dependent on CD11b/CD18 and ICAM-1,[152] and anti-CD11a monoclonal antibodies have little effect on this adhesion. In contrast, ICAM-1-dependent adhesion of canine neutrophils to cytokine-stimulated canine jugular vein endothelial cells is largely dependent on CD11a/CD18 with some contribution of CD11b/CD18. Furthermore, the biologic consequences of these interactions appear to be distinct. Upon contacting cytokine-stimulated endothelial monolayers, a high percentage of canine neutrophils rapidly migrate through the monolayers,[62] but upon contacting the cytokine-stimulated myocytes, neutrophils adhere, show little tendency to migrate, and within minutes begin to secrete reactive oxygen species.[152] Adherence-dependent hydrogen peroxide production[153,154] can be mediated by CD11b/CD18,[130] but evidence is lacking that CD11a/CD18 can trigger the oxidative burst in neutrophils.[130]

ELAM-1. Smith et al.[24] found that the adhesiveness of HUVEC monolayers for CD18-deficient neutrophils peaked at four hours after IL-1 stimulation and declined thereafter to near baseline values within eight hours. This time course is consistent with that of endothelial surface expression of ELAM-1,[112] and Luscinskas et al.[31,145] published evidence from monoclonal antibody blocking studies that ELAM-1-dependent and CD18-ICAM-1-dependent adherence mechanisms are distinct. Smith et al.[24] also found that transendothelial migration of CD18-deficient neutrophils through cytokine-stimulated endothelial monolayers at the peak time of adhesion was extremely low,[24,56] in marked contrast to normal neutrophils. These results suggest that while CD18 integrins may be necessary for transmigration, ELAM-1-dependent adhesion is not sufficient for migration. Indeed, patients with CD18-deficiency have a profound deficit in migration of neutrophils into sites of inflammation, with possibly an important exception. Hawkins et al.[155] observed in one patient the accumulation of neutrophils in the alveolar spaces in association with gram-positive infection of the lungs. Doerschuk et al.[156] found that systemic administration of anti-CD18 monoclonal antibody, 60.3, almost completely inhibited emigration of neutrophils into the lungs of rabbits in response to phorbol ester, *Escherichia coli* organisms, and *E coli* endotoxin, but was without effect on the response to *Streptococcus pneumoniae.* In contrast to the results in lung, 60.3 largely prevented the emigration of neutrophils into the peritoneal cavity of rabbits in response to *S pneumoniae*, indicating that mechanisms of extravasation may vary with different vascular beds. These studies show that transendothelial migration *in vivo* can be CD18-independent, but whether ELAM-1 is the predominant determinant under these conditions has not been assessed. Also, no published studies *in vitro* have revealed an ELAM-1-dependent, CD18-independent mechanism for transmigration of neutrophils.

Evaluations of the participation of ELAM-1 in neutrophil migration *in vitro* have provided inconsistent results. Luscinskas et al.[31] found that transendothelial migration peaked following IL-1 stimulation of HUVECs at a time when surface expression of ELAM-1 was highest (i.e., four hours), and that anti-ELAM-1 monoclonal antibody, H18/7, almost completely inhibited migration. Anti-ICAM-1 monoclonal antibody, Hu5/3, was also markedly inhibitory in their model, supporting the conclusion that both ICAM-1 and ELAM-1 were necessary for migration. Kishimoto et al.,[63] using the *in vitro* model of transendothelial migration reported by Smith et al.,[24] found that anti-ELAM-1 monoclonal antibodies, CL2 and CL3, had little effect on migration. In both investigations, adhesion of isolated, unstimulated neutrophils to cytokine-stimulated HUVECs was partially inhibited by either anti-ICAM-1 or anti-ELAM-1 monoclonal antibodies, and when these antibodies were combined in the same assay, inhibition was additive. In light of the observations *in vivo* suggestive of varying mechanisms for neutrophil emigration in different vascular beds, the apparent inconsistency *in vitro* may reflect differing requirements for ELAM-1 in neutrophil transmigration. One technical difference between these studies that may be significant was that in the Luscinskas et al. study, endothelial cell monolayers were prepared on type I collagen gels, and in the Kishimoto et al. study, they were prepared on fibronectin-coated glass. Because the activation of neutrophil motility in these *in vitro* models is endothelial-dependent, investigation of possible differential influences of the substrates on the relevant endothelial functions may prove important. Furie et al.[157] have recently evaluated the effects of anti-ELAM-1 monoclonal antibodies BB11[158] and H18/7[31] on neutrophil migration through endothelial monolayers grown on amnion.[30] They found that when the endothelium was stimulated with low concentrations of rIL-1β (0.1 to 0.15 U/ml), anti-ELAM-1 antibodies were partially effective, but when the IL-1 level was optimal for provoking transendothelial migration, these antibodies were ineffective in contrast to anti-CD18, which profoundly inhibited migration. They concluded that ELAM-1 was not necessary for migration in this assay.

As discussed above, the mechanism(s) by which neutrophil motility is activated by cytokine-stimulated endothelial cells is unknown but may involve endothelial-derived chemotactic factors such as PAF and IL-8. Lo et al.[159] have presented evidence for an additional determinant. They found that coincubating neutrophils, iC3b-coated erythrocytes, and cytokine-stimulated HUVECs on fibronectin-coated plastic for 20 minutes at 37°C resulted in a high degree of erythrocyte binding to the neutrophils that was inhibited by preincubation of the HUVECs with anti-ELAM-1 monoclonal antibodies, BB11 or H18/7, but not by anti-ICAM-1 monoclonal antibody LB-2 (an antibody that inhibits only LFA-1–ICAM-1 interactions[149]). Activation of iC3b binding also occurred when neutrophils were allowed to contact for one hour at 37°C soluble recombinant ELAM-1 adsorbed to a plastic surface. Anti-ELAM-1 antibodies reduced the number of erythrocytes bound per neutrophil to baseline levels (i.e., when neutrophils contacted glycophorin or human serum albumin-coated plastic). In addition, the soluble recombinant ELAM-1 stimulated neutrophil motility. These authors speculated that ELAM-1 may provoke motility of interacting neutrophils and increase CD18-dependent adhesion to endothelial cells. If such a mechanism primarily determined the transmigration seen in the Luscinskas et al. study,[31] then both ELAM-1 and ICAM-1 would be necessary for migration. However, if there is redundancy in the endothelial mechanisms that stimulate neutrophil motility (e.g., IL-8 and PAF also functioning in this role), then anti-ELAM-1 antibodies may not inhibit migration of adherent neutrophils.

CONTRIBUTION OF MARGINATION TO NEUTROPHIL LOCALIZATION

Intravital microscopy has clearly revealed that neutrophils can roll along the vessel wall in postcapillary venules[160–164] and that this phenomenon is markedly increased in the initial stages of acute inflammation.[48,165–167] The rolling cells often stop, change shape, and emigrate into surrounding tissue. Arfors et al.[168] initially raised the possibility that the molecular determinants of the rolling phenomenon were distinct from the stopping and transmigration phenomena in a study on the effects of anti-CD18 monoclonal antibody, 60.3, in a rabbit model of inflammation. They observed that systemic administration of this antibody almost completely blocked the stopping and transmigration of neutrophils, but the rolling phenomenon appeared to be unaffected. Lawrence et al.[136,169] used a parallel plate flow chamber to assess the ability of isolated neutrophils to adhere while flowing past a confluent monolayer of HUVECs. Without stimulation, few neutrophils were found to interact with the HUVEC monolayer. However, when the endothelial cells were stimulated with IL-1 for three to four hours prior to the flow experiment, a marked increase in adhesion was seen at wall shear stresses below 3 dynes/cm^2. Consistent with observations *in vivo,* two types of adhesion were observed, rolling and stationary. At a wall shear stress of approximately 2 dynes/cm^2, rolling cells were usually moving at less than 10 µm/sec, in contrast to cells that did not interact with the endothelial surface, where the rate of movement was approximately 2000 µm/sec. Stationary cells often changed shape and migrated beneath the monolayer as previously seen[24] (see Figure 5-1). Abbassi et al.[62] have confirmed these observations using canine neutrophils and canine jugular vein endothelial cell monolayers. In both the Lawrence et al. and Abbassi et al. studies, anti-CD18 monoclonal antibodies were without effect on the number of neutrophils associating with the cytokine-stimulated monolayers at wall shear stresses around 2 dynes/cm^2, but there was a pronounced effect on the behavior of the neutrophils in the presence of the antibody. Transendothelial migration was markedly reduced, fewer neutrophils stopped on the apical surface of the endothelium, and the rate of rolling was significantly increased. Lawrence et al.[136,169] observed that CD18-deficient neutrophils associated with the monolayers as well as normal cells, but their rate of rolling was increased and transmigration did not occur. Perry and Granger[170] found that the rolling velocity of neutrophils in the cat mesenteric venules was increased by administration of anti-CD18 monoclonal antibody, IB$_4$, a finding that supports previous predictions that the velocity of rolling is influenced by neutrophil–endothelial adherence. Lawrence et al.[136,169] also found that anti-ICAM-1 antibody, R6.5, did not reduce the number of monolayer-associated neutrophils, but greatly inhibited transmigration. These studies indicate that both *in vitro* and *in vivo*, the adherence mechanisms of CD18–ICAM-1 do not account for the initial rolling (i.e., margination), but they apparently play critical roles in events subsequent to margination.

LECAM-1 (LAM-1). Lectin adhesion molecule-1 (LECAM-1), also known as leukocyte adhesion molecule-1 (LAM-1), was originally identified in the mouse by monoclonal antibody MEL-14, and numerous monoclonal antibodies are now available against the human homologue.[171,172] On lymphocytes,

LECAM-1 functions as the peripheral lymph node homing receptor and binds through its lectin domain to a carbohydrate-containing ligand on high endothelial venules in peripheral lymph nodes.[173] Smith et al.[64] and Abbassi et al.[62] found that anti-LECAM-1 monoclonal antibodies would reduce by greater than 60% the number of neutrophils associating with cytokine-stimulated endothelial monolayers under conditions of flow described above.[136,169] Additionally, antibodies to LECAM-1 did not reduce the ability of adherent neutrophils to migrate through cytokine-stimulated endothelial monolayers *in vitro*,[62,64,138] suggesting that the principal role for this glycoprotein on neutrophils is in margination. The topography of LECAM-1 may be critical to this function. A recent report shows LECAM-1 to be concentrated on the surface of microvillus-like projections from unstimulated neutrophils.[174] In contrast, CD18 appeared on flat areas of the cell surface as well as on the surface of projections and was abundant on the membranes of cytoplasmic granules.[87]

Von Andrian et al.[175] reported that systemic administration of anti-LECAM-1 monoclonal antibody, DREG-200, in rabbits greatly reduced the number of rolling cells *in vivo,* and Ley et al.[176] reported that systemic administration in rats of a soluble recombinant form of LECAM-1 or a polyclonal anti-LECAM-1 antiserum also markedly reduced by greater than 80% the number of rolling cells *in vivo*. The soluble form of LECAM-1 presumably blocked the ligand on activated endothelial cells, making it inaccessible to neutrophils. Thus, it appears that both *in vitro* and *in vivo,* the constitutively expressed neutrophil glycoprotein, LECAM-1, interacts with a newly upregulated structure on the surface of endothelial cells and accounts for a significant portion of the initial margination. Studies in the flow chamber *in vitro*[62,64,138] demonstrated that reducing the number of neutrophils marginating on endothelial monolayers (e.g., with anti-LECAM-1 monoclonal antibody) reduced transendothelial migration simply by limiting the number of neutrophils available to migrate. This mechanism may explain the anti-inflammatory effects of anti-LECAM-1 monoclonal antibody, MEL-14, seen by Lewinsohn et al.[177] and Jutila et al.[140] in mice, and the anti-inflammatory effects of the soluble LECAM-1[178] seen by Watson et al.

LECAM-1-dependent neutrophil–endothelial adhesion is lost following chemotactic stimulation of the neutrophil,[64,179] an effect most clearly seen under conditions of flow where the contributions of the newly-activated CD18 heterodimers do not obscure the loss of LECAM-1-dependent adherence.[62,64,138] The most likely explanation for this phenomenon is that stimulation results in shedding of LECAM-1 from the cell surface.[39,62,64,140,140a,180] There has been a high degree of correlation observed between the amount of LECAM-1 on the neutrophil surface and adhesion to cytokine-stimulated endothelial monolayers under conditions of flow *in vitro*,[64,138] and Jutila et al.[140] found that isolated murine neutrophils had reduced ability to migrate into thioglycollate-induced peritonitis if they had been activated *in vitro* to shed LECAM-1, labelled, and returned to the animal intravenously, though a high percentage continued to circulate throughout the time course of the inflammation. Thus, the loss of LECAM-1 function (either by shedding or inhibition with monoclonal antibody) may reduce the ability of neutrophils to localize at sites of inflammation, probably by reducing margination on inflamed endothelium.

ELAM-1. Margination *in vitro* does not occur at venous wall shear stresses unless the endothelial cells have been stimulated.[136] Conditions that lead to upregulation of ELAM-1 lead to pronounced margination, thus supporting speculation that ELAM-1 could play an important role. However, anti-LECAM-1 monoclonal antibodies produce between 60% and 80% inhibition of neutrophil margination both *in vitro* and *in vivo*[64,138,175] and greater than 60% inhibition of neutrophil accumulation in peritoneal inflammation, leaving the impression that ELAM-1 may not be sufficient to sustain but a minor part, if any, of the margination phenomenon. Alternatively, ELAM-1 may play a major role by functioning as a necessary accessory molecule along with LECAM-1. Recent studies by Mulligan et al.[181] have shown that anti-ELAM-1 monoclonal antibody, CL3 (originally made against human ELAM-1),[63,148] cross-reacts with ELAM-1 in the rat. When administered systemically, CL3 significantly inhibited neutrophil influx into glycogen-induced peritonitis, and immune complex-induced inflammation in the skin and lungs. Gundel et al.[182] found that anti-human ELAM-1 monoclonal antibody CL2[63,148] was cross-reactive with cynomolgus ELAM-1. They found that monkeys sensitive to *Ascaris suum* extract expressed ELAM-1 on lung vascular endothelium within six hours following a single inhalation exposure to the antigen, and that this correlated with influx of neutrophils into the lungs and late-phase airway obstruction. Systemic administration of CL2 blocked both influx of neutrophils and the airway obstruction. Though these observations do not reveal whether ELAM-1 is primarily involved in margination or transmigration, they indicate an important role for ELAM-1 in acute inflammation at these sites.

As discussed above, studies of transendothelial migration *in vitro* have not resolved a function for ELAM-1, but recent studies support the idea that ELAM-1 is involved in the same adherence pathway as neutrophil LECAM-1. Kishimoto et al.[63] initially observed that the inhibitory effects of anti-LECAM-1 and anti-ELAM-1 monoclonal antibodies were not additive, but each produced additive inhibition when combined with anti-ICAM-1. Neither anti-LECAM-1 nor anti-ELAM-1 were inhibitory when neutrophil adhesion to endothelial monolayers was primarily determined by ICAM-1, that is, at more than eight hours after stimulation of the endothelial cells with IL-1.[24,31] Chemotactic stimulation of neutrophils under conditions shown to reduce neutrophil LECAM-1 to very low levels markedly inhibited ELAM-1-dependent neutrophil adhesion, an observation consistent with the results of Dobrina et al.[183] Furthermore, anti-LECAM-1 significantly inhibited neutrophil adhesion to COS cells and L cells transfected with ELAM-1. Because these studies were carried out under static conditions, more recent work has assessed neutrophil adhesion under conditions of flow (Abbassi, Kishimoto, and Smith, unpublished observations). Using ELAM-1-transfected L cell monolayers in the flow chamber as previously described,[64] neutrophils were found to exhibit a rolling type adhesion that was almost completely inhibited by anti-ELAM-1 monoclonal antibody, CL3, and was inhibited by greater than 65% by anti-LECAM-1 antibodies. Chemotactic stimulation shown to markedly deplete neutrophil surface LECAM-1 reduced adhesion to the same extent as anti-LECAM-1.

The speculation that ELAM-1 and neutrophil LECAM-1 may utilize the same adhesion pathway during margination was strengthened by the recent studies of Picker et al.[174] They found that the carbohydrate moiety recognized by the lectin domain of ELAM-1, sialyl Lewis-X,[184-186] is present on LECAM-1 isolated from neutrophils but not on LECAM-1 isolated from lymphocytes. Furthermore, mouse pre-B cells, L1-2, transfected with ELAM-1, adhered to surfaces with adsorbed neutrophil LECAM-1 but not surfaces with lymphocyte LECAM-1. This adherence was inhibited by antibodies to ELAM-1, sialyl Lewis-X or LECAM-1. Thus, it appears that neutrophil LECAM-1 is modified with ELAM-1-binding oligosaccharide structures that are not present on lymphocyte LECAM-1, and these structures support ELAM-1-dependent neutrophil margination on IL-1-stimulated endothelial cells under conditions of flow.

GMP-140 (CD62). A third member of the lectin, epidermal growth factor-complement binding cell adhesion molecule (LECCAM) (selectin) family appears to be involved in margination and possibly activation of neutrophils at the endothelial surface. CD62 is constitutively expressed in normal, non-inflamed endothelium[187] and is found within the Weibel-Palade bodies.[187-189] It is rapidly translocated to the apical surface of the cell following stimulation with such agents as thrombin or histamine,[187,189,190] where it remains for less than one hour.[189] Patel et al.[191] found that surface expression of CD62 was markedly prolonged when low level oxidants were the stimuli for translocation from the Weibel-Palade bodies. Geng et al.[190] and Patel et al.[191] found that CD62 can under static conditions attach neutrophils to endothelial cells. Lawrence and Springer[193] demonstrated that CD62 in artificial planar membranes will cause neutrophils to roll under conditions of flow at wall shear stresses similar to those demonstrated by Smith et al.[64] and Abbassi et al.[62] to cause LECAM-1-dependent rolling on endothelial cells. Thus, it appears that members of the LECCAM (selectin) family have a high enough rate of reaction with their carbohydrate epitopes to catch neutrophils as they flow by the endothelium.

In addition to the possible role of CD62 in promoting margination, Lorant et al.[192] have obtained evidence that binding of CD62 with its ligand on neutrophils potentiates CD11/CD18 adhesive responses stimulated by agonists such as chemotactic factors. This would result in strengthening of adhesion and possibly promote more effective leukocyte migration.

THREE STAGES IN THE PROCESS OF LEUKOCYTE EXTRAVASATION

Figure 5-3 illustrates schematically three stages in the extravasation of neutrophils from postcapillary venules in which adhesion molecules appear to play important roles. The rolling adhesion often seen early during inflammation involves members of the LECCAM (selectin) family. Because it is constitutively expressed and rapidly mobilized, CD62 may be in position to catch previously unstimulated neutrophils at very early stages. If the inflammation involves cytokine production (e.g., TNF), upregulation of ELAM-1 and possibly ligands for LECAM-1 would serve to sustain the phenomenon of margination. While the velocity of rolling appears to be influenced by CD18 integrins,[62,136,190,193] they do not appear to be capable of independently sustaining the rolling phenomenon,[62,64,136,193] even at venous wall shear stresses.

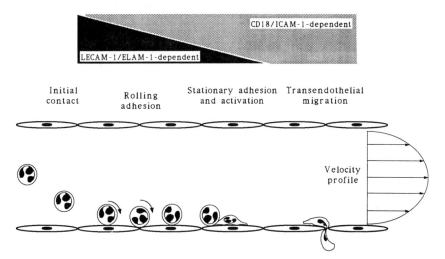

Figure 5-3. Schematic representation of three stages in the process of leukocyte extravasation from a postcapillary venule. Illustrated are rolling adhesion, stationary adhesion, and migration. The theoretical relative contributions of the LECCAM (selectin) and CD18/ICAM-1 adhesion pathways are illustrated symbolically.

Stationary adhesion is associated with obvious activation of the neutrophil. The cells change shape, becoming ruffled, flattened, and bipolar. The mechanisms that account for these events are not well-defined, though locally generated chemotactic factors probably contribute, and the adhesive mechanisms that initially catch the neutrophils may also contribute.[159,192] This stage accomplishes an important transition for the neutrophil, not only from an unactivated to a motile cell, but from a primarily LECCAM-dependent adhesion to predominantly CD18-dependent adhesion.[140,141] Downregulation of neutrophil LECAM-1 and upregulation of neutrophil CD11b/CD18 occur upon contact of neutrophils with cytokine-stimulated endothelial cells *in vitro*,[64] and neutrophils have markedly reduced surface LECAM-1 after diapedesis *in vitro*[52a] and *in vivo*.[140,140a] Transendothelial migration of neutrophils in most of the settings evaluated requires CD18 integrins. Though some evidence indicates that CD18-independent neutrophil extravasation can occur in vivo, morphologic and molecular mechanisms remain obscure.

This model is consistent with the bulk of evidence obtained *in vitro* and *in vivo*. It is important to note, however, that the experimental settings *in vitro* do not fully reproduce the complex conditions *in vivo,* and the relative contributions of each adhesive mechanism are likely to vary with different vascular beds and inflammatory etiologies. If, for example, neutrophil influx occurs through smaller vessels than postcapillary venules, then rolling-type margination may not occur. Rather, physical factors may determine the initial stationary phase of localization prior to neutrophil activation, or systemically-activated neutrophils may lodge in small diameter vessels.[194] Evidence regarding a role for LECCAMs *in vivo* is much more limited than for CD18, and at present there is no published evidence supporting a role for GMP-140 in neutrophil localization *in vivo*.

SPECIFIC ADHESION MOLECULES AND THE MIGRATION OF LYMPHOCYTES AND OTHER LEUKOCYTES

Lymphocytes. Kavanaugh et al.[124] and Oppenheimer-Marks et al.[125] investigated the contribution of specific adhesion molecules to transendothelial migration of T cells *in vitro*. The assay used in these studies involved HUVECs grown to confluence on micropore nitrocellulose filters.[195] This assay was shown previously to reflect spontaneous lymphocyte migration, migration toward a chemotactic gradient, or migration provoked by cytokine stimulation of the endothelium. An important contribution of CD18 integrins was demonstrated by the finding that T cell clones from patients with CD18 deficiency migrated at about 50% of the level of normal T cells. CD11a/CD18 appears to be an important integrin in this process, as monoclonal antibodies to

CD11a and CD54 (ICAM-1) produced significant inhibition of migration of normal lymphocytes. Van Epps et al.[196] found that IL-1β-stimulation of HUVEC monolayers promoted T-cell transmigration, and this migration was markedly reduced by antibodies against CD18 and CD11a.

Analogous to neutrophil–endothelial interactions, T cell–endothelial interactions are modulated by the state of activation of each cell type.[197] The contributions of CD11a/CD18 to adhesion are greatly increased by stimulation of the lymphocytes. This appears to result from a qualitative change in surface CD11a/CD18. Phorbol esters stimulate CD11a/CD18-dependent aggregation but do not alter the surface density of CD11a/CD18,[117-120] a phenomenon linked to an apparent change in avidity of CD11a/CD18 binding to ICAM-1.[116,122,123] This can be most readily seen on planar membranes containing ICAM-1. Such stimuli as cross-linking the T cell receptor[116] will transiently regulate the adhesion of T cells within a period of 30 minutes without altering surface density of CD11a/CD18. This qualitative change in CD11a/CD18 is recognized by two monoclonal antibodies. One (monoclonal antibody NKI-L16) binds to an epitope that induces homotypic aggregation.[198] This epitope is not present on unstimulated lymphocytes and depends on Ca++ for its activation.[123] The second antibody (monoclonal antibody 24) recognizes a Mg++-dependent epitope that is only present after activation of the cell.[106] It remains an intriguing hypothesis that such transient changes in the avidity of CD11a/CD18 for its ligands (e.g., ICAM-1) provide one mechanism for adherence-dependent locomotion of the T cell.

Stimulation of the endothelial cell (e.g., with IL-1β) increases surface expression of not only ICAM-1 and ELAM-1, but vascular cell adhesion molecule-1 (VCAM-1). Oppenheimer-Marks et al.[125] found that this change lead to a predominant role for VCAM-1 and VLA-4 (CD49d/CD29). While this receptor–ligand pair was important to adhesion, these investigators were unable to find a contribution of VCAM-1 to transendothelial migration. In contrast, ICAM-1 played a prominent role in migration. Of particular interest was their finding that VCAM-1 was limited to the apical surface of the endothelial monolayer while ICAM-1 was prominent both on the basal and apical sides of the endothelial cells, and it was consistently present at intercellular junctions. Thus, it appears that one set of adhesion molecules (VLA-4 on the lymphocyte and VCAM-1 on the endothelial cell) contributes to adhesion while another (CD11a/CD18 and ICAM-1) contributes to transmigration, at least in part because of the surface distribution of the VCAM-1 and ICAM-1 on activated endothelium.

A subset of T cells (resting memory cells) binds ELAM-1.[148,197,199,200] As appears to be the case with neutrophils, ELAM-1 may serve to catch lymphocytes as they flow past sites of inflammation. Immunohistology has demonstrated ELAM-1 at chronic inflammatory sites where neutrophil influx is minimal or absent,[148] and there ELAM-1 may play a role in the preferential localization of memory cells. The role of ELAM-1 in transendothelial migration of T cells is unknown, but may serve only to adhere these cells to endothelium. Extravasation may be dependent on the local production of activation or chemotactic factors such as IL-8.[201]

In addition to emigration at sites of inflammation, lymphocytes migrate out of the vascular system after adhering to high endothelial venules. This phenomenon has been extensively studied, and only one hypothesis possibly related to transendothelial migration will be discussed here. In addition to a potential role for CD11a/CD18-ICAM-1 as indicated above, another adhesion molecule, LECAM-1, is well known to account for adherence to high endothelial venules.[173] In a role possibly analogous to CD11a/CD18, LECAM-1 appears to undergo an activation-induced, transient increase in affinity for specific carbohydrate ligands. Spertini et al.[202] found that stimuli such as cross-linking the T cell receptor would markedly increase the binding of polyphosphomannan ester core polysaccharide derived from the yeast *Hansenula* (PPME) without increasing the density of LECAM-1 on the leukocyte surface. This increased binding of PPME was blocked by monoclonal antibodies against LECAM-1. These authors speculated that sequential activation and deactivation of adhesion molecules like LECAM-1, CD11a/CD18, and VLA-4[116,203] may contribute to lymphocyte migration.

Eosinophils. These leukocytes show little ability to adhere to resting endothelium *in vitro*, but following cytokine stimulation, the endothelial cells become adhesive for both eosinophils and basophils.[204-206] ELAM-1 and ICAM-1 contribute to the adherence,[207] but in contrast to neutrophils, both eosinophils and basophils express VLA-4 (CD49d/CD29) and thereby have the ability to bind VCAM-1.[204,208] Current evidence indicates that VCAM-1 is a major ligand for eosinophil adhesion to activated endothelium,[204] and may represent a means by which recruitment of this cell type is distinct from neutrophils (because neutrophils do not bind VCAM-1). A role for VCAM-1 in transendothelial migration of eosinophils is not evident, though, and if as observed by Oppenheimer-Marks et al.,[125] VCAM-1 distribution is limited to the luminal surface of activated endothelial cells, it is likely that the phenomenon proposed for neutrophils would hold true for eosinophils—extravasation depends on a second, distinct set of mechanisms. Wegner et al.[206] found that eosinophil infiltration in lungs of *Ascaris* extract-sensitized cynomolgus monkeys was significantly reduced by systemic administration of anti-ICAM-1 monoclonal antibody, and the airway hyper-responsiveness was significantly reduced. These results may indicate that ICAM-1 is needed for transendothelial migration of eosinophils. ICAM-1-dependent adherence of eosinophils to endothelium *in vitro* is enhanced by chemotactic stimulation.[206] The relative contributions of VCAM-1 and ELAM-1 to eosinophil localization *in vivo* remain unknown.

REFERENCES

1. Grant, L. (1974) *The Inflammatory Process* (Zweifach, B.W., Grant, L., and McCluskey, R.T., eds.) pp. 205–221, Academic Press, New York.
2. Hurley, J.V. (1963) *Austral. J. Exp. Biol. Med. Sci.* 41, 171–186.
3. Marchesi, V.T. (1964) *Ann. N.Y. Acad. Sci.* 116, 774–788.
4. Marchesi, V.T., and Gowans, J.L. (1964) *Proc. Royal Soc. Med.* 159, 283–290.
5. Shaw, J.O (1980) *Am. J. Pathol.* 101, 283–302.
6. Beesley, J.E., Pearson, J.D., Hutchings, A., Carleton, J.S., and Gordon, J.L. (1979) *J. Cell Sci.* 38, 237–248.
7. Beesley, J.E., Pearson, J.D., Carleton, J.S., Hutchings, S.A., and Gordon, J.L. (1978) *J. Cell Sci.* 33, 85–101.
8. Lackie, J.M. and Debono, D. (1977) *Microvasc. Res.* 13, 107–112.
9. Taylor, R.F., Price, T.H., Schwartz, S.M., and Dale, D.C. (1981) *J. Clin. Invest.* 67, 584–587.
10. Furie, M.B., Cramer, E.V., Naprstek, B.L., and Silverstein, S.C. (1984) *J. Cell Biol.* 98, 1033–1041.
11. Furie, M.B., Naprstek, B.L., and Silverstein, S.C. (1987) *J. Cell Science* 88, 161–175.
12. Granstein, R.D., Margolis, R.J., Mizel, S.D., and Sauder, D.N. (1986) *J. Clin. Invest.* 77, 1020–1027.
13. Williamson, J.R. and Grisham, J.W. (1961) *Am. J. Pathol.* 39, 239–256.
14. Huber, A.R. and Weiss, S.J. (1989) *J. Clin. Invest.* 83, 1122–1136.
15. Beck, G., Habicht, G.S., Benach, J.L., and Miller, F. (1986) *J. Immunol.* 136, 3025–3032.
16. Cybulsky, M.I., McComb, D.J., and Movat, H.Z. (1988) *J. Immunol.* 140, 3144–3149.
17. Henderson, B. and Pettipher, E.R. (1988) *Biochem. Pharm.* 37, 4171.
18. Movat, H.Z., Burrowes, C.E., Cybulsky, M.I., and Dinarello, C.A. (1987) *Am. J. Pathol.* 129, 463–476.
19. Rampart, M. and Williams, T.J. (1988) *Br. J. Pharmacol.* 94, 1143–1151.
20. Sayers, T.J., Wiltrout, T.A., Bull, C.A., Denn, A.C., Pilaro, A.M., and Lokesh, B. (1988) *J. Immunol.* 141, 1670–1677.
21. Sharpe, R.J., Margolis, R.J., Askari, M., Amento, E.P., and Granstein, R.D. (1988) *J. Invest. Dermatol.* 91, 353–357.
22. Wankowicz, Z., Megyeri, P., and Issekutz, A. (1988) *J. Leuk. Biol.* 43, 349–356.
23. Pober, J.S., and Cotran, R.S. (1990) *Transplantation* 50, 537–544.
24. Smith, C.W., Rothlein, R., Hughes, B.J., Mariscalco, M.M., Schmalstieg, F.C., and Anderson, D.C. (1988) *J. Clin. Invest.* 82, 1746–1756.
25. Armstrong, P.B. and Lackie, J.M. (1975) *J. Cell Biol.* 65, 439–462.
26. Buchanan, M.R., Crowley, C.A., Rosin, R.E., Gimbrone, Jr., M.A., and Babior, B.M. (1982) *Blood* 60, 160–165.
27. Gimbrone, Jr., M.A., Brock, A.F., and Schafer, A.I. (1984) *J. Clin. Invest.* 74, 1552–1555.
28. Tonnesen, M.G., Smedly, L.A., and Henson, P.M. (1984) *J. Clin. Invest.* 74, 1581–1592.

29. Moser, R., Schleiffenbaum, B., Groscurth, P., and Fehr, J. (1989) *J. Clin. Invest.* 83, 444–455.
30. Furie, M.B. and McHugh, D.D. (1989) *J. Immunology* 143, 3309–3317.
31. Luscinskas, F.W., Cybulsky, M.I., Kiely, J.M., Peckins, C.S., Davis, V.M., and Gimbrone, M.A. (1991) *J. Immunol.* 146, 1617–1625.
32. Kharazmi, A., Nielsen, H., and Bendzen, K. (1988) *Immunobiology* 177, 32–37.
33. Yoshimura, T., Matsushima, K., Oppenheim, J.J., and Leonard, E.J. (1987) *J. Immunol.* 139, 788–793.
34. Breviario, F., Bertocchi, F., Dejana, E., and Bussolino, F. (1988) *J. Immunol.* 141, 3391–3397.
35. Bussolino, F., Camussi, G., and Baglioni, C. (1988) *J. Biol. Chem.* 263, 11856–11861.
36. Farber, H.W., Center, D.M., and Rounds, S. (1985) *Circ. Res.* 57, 898–903.
37. Farber, H.W., Weller, P.F., Rounds, S., Beer, D.J., and Center, D.M. (1986) *J. Immunol.* 137, 2918–2923.
38. Gimbrone, Jr., M.A., Obin, M.S., Brock, A.F., Luis, E.A., Hass, P.E., Hebert, C.A., Yip, Y.K., Leung, D.W., Lowe, D.G., Kohr, W.J., Darbonne, W.C., Bechtol, K.B., and Baker, J.B. (1989) *Science* 246, 1601–1603.
39. Griffin, J.D., Spertini, O., Ernst, T.J., Belvin, M.P., Levine, H.B., Kanakura, Y., and Tedder, T.F. (1990) *J. Immunol.* 145, 576–584.
40. Gudewicz, P.W., Odekon, L.E., DelVecchio, P.J., and Saba, T.M. (1988) *J. Leuk. Biol.* 440, 1–7.
41. Mercandetti, A.J., Lane, T.A., and Colmerauer, M.E.M. (1984) *J. Lab. Clin. Med.* 104, 370–377.
42. O'Brien, R.F., Seton, M.P., Makarski, J.S., Center, D.M., and Rounds, S. (1984) *Am. Rev. Respir. Dis.* 130, 103–109.
43. Schroder, J.M. and Christophers, E. (1989) *J. Immunol.* 142, 244–251.
44. Strieter, R.M., Kunkel, S.L., Showell, H.J., Remick, D.G., Phan, S.H., Ward, P.A., and Marks, R.M. (1989) *Science* 243, 1467–1469.
45. Zimmerman, G.A., McIntyre, T.M., Mehra, M., and Prescott, S.M. (1990) *J. Cell Biol.* 110, 529–540.
46. Zavoico, G.B., Ewenstein, B.M., Schafer, A.I., and Pober, J.S. (1989) *J. Immunol.* 142, 3993–3999.
47. Goetzl, E.J. and Pickett, W.C. (1980) *J. Immunol.* 125, 1789–1794.
48. Kubes, P., Ibbotson, G., Russell, J., Wallace, J.L., and Granger, D.N. (1990) *Am. J. Physiol.* 259, G300–G305.
49. Baggiolini, M., Walz, A., and Kunkel, S.L. (1989) *J. Clin. Invest.* 84, 1045–1049.
50. Detmers, P.A., Lo, S.K., Olsen-Egbert, E., Walz, A., Baggiolini, M., and Cohn, Z.A. (1990) *J. Exp. Med.* 171, 1155–1162.
51. Hechtman, D.H., Cybulsky, M.I., Fuchs, H.J., Baker, J.B., and Gimbrone, Jr., M.A. (1991) *J. Immunol.* 147, 883–892.
52. Matsushima, K. and Oppenheim, J.J. (1989) *Cytokine* 1, 2–10.
52a. Huber, A.R., Kunkel, S.L., Todd, R.F., and Weiss, S.J. (1991) *Science* 254, 99–102.
53. Charo, I.F., Yuen, C., Perez, H.D., and Goldstein, I.M. (1986) *J. Immunol.* 136, 3412–3419.
54. Gamble, J.R., Harlan, J.M., Klebanoff, S.J., and Vadas, M.A. (1985) *Proc. Natl. Acad. Sci. USA* 82, 8667–8674.

55. Pohlman, T.H., Stanness, K.A., Beatty, P.G., Ochs, H.D., and Harlan, J.M. (1986) *J. Immunol.* 136, 4548–4553.
56. Smith, C.W., Marlin, S.D., Rothlein, R., Toman, C., and Anderson, D.C. (1989) *J. Clin. Invest.* 83, 2008–2017.
57. Tonnesen, M.G., Anderson, D.C., Springer, T.A., Knedler, A., Avdi, N., and Henson, P.M. (1989) *J. Clin. Invest.* 83, 637–646.
58. Zimmerman, G.A. and McIntyre, T.M. (1988) *J. Clin. Invest.* 81, 531–537.
59. Entman, M.L., Youker, K., Shappell, S.B., Siegel, C., Rothlein, R., Dreyer, W.J., Schmalstieg, F.C., and Smith, C.W. (1990) *J. Clin. Invest.* 85, 1497–1506.
60. Carveth, H.J., Bohnsack, J.F., McIntyre, T.M., Baggiolini, M., Prescott, S.M., and Zimmerman, G.A. (1989) *Biochem. Biophys.Res.Commun.* 162, 387–393.
61. Hechtman, D.H., Cybulsky, M.I., Baker, J.B., and Gimbrone, M.A. (1990) *FASEB J.* 4, A890. Abstract.
62. Abbassi, O., Lane, C.L., Krater, S.S., Kishimoto, T.K., Anderson, D.C., McIntire, L.V., and Smith, C.W. (1991) *J. Immunol.* 147, 2107–2115.
63. Kishimoto, T.K., Warnock, R.A., Jutila, M.A., Butcher, E.C., Lane, C.L., Anderson, D.C., and Smith, C.W. (1991) *Blood* 78, 805–811.
64. Smith, C.W., Kishimoto, T.K., Abbassi, O., Hughes, B.J., Rothlein, R., McIntire, L.V., Butcher, E., and Anderson, D.C. (1991) *J. Clin. Invest.* 87, 609–618.
65. Lo, S.K., Detmer, P.A., Levin, S.M., and Wright, S.D. (1989) *J. Exp. Med.* 169, 1779–1793.
66. Zigmond, S.H. (1977) *J.Cell Biol.* 75, 606–616.
67. Zigmond, S.H. (1978) *J.Cell Biol.* 77, 269–287.
68. Abramson, J.S., Mills, E.L., Sawyer, M.K., Regelmann, W.R., Nelson, J.D., and Quie, P.G. (1981) *J. Pediatr.* 99, 887–894.
69. Anderson, D.C., Schmalstieg, F.C., Kohl, S., Arnaout, M.A., Hughes, B.J., Tosi, M.F., Buffone, G.J., Brinkley, B.R., Dickey, W.D., Abramson, J.S., Springer, T.A., Boxer, L.A., Hollers, J.M., and Smith, C.W. (1984) *J. Clin. Invest.* 74, 536–551.
70. Arnaout, M.A., Pitt, J., Cohen, H.J., Melamed, J., Rosen, F.S., and Colten, H.R. (1982) *N. Engl. J. Med.* 306, 693–699.
71. Bissenden, J.G., Haeney, M.R., Tarlow, M.J., and Thompson, R.A. (1981) *Arch. Dis. Child.* 56, 397–399.
72. Bowen, T.J., Ochs, H.D., Altman, L.C., Price, T.H., Van Epps, D.E., Brautigan, D.L., Rosin, R.E., Perkins, W.D., Babior, B.M., Klebanoff, S.J., and Wedgwood, R.J. (1982) *J. Pediatr.* 101, 932–940.
73. Crowley, C.A., Curnutte, J.T., Rosin, R.E., Andre-Schwartz, J., Gallin, J.I., Klempner, M., Snyderman, R., Southwick, F.S., Stossel, T.P., and Babior, B.M. (1980) *N. Engl. J. Med.* 302, 1163–1168.
74. Davies, E.G., Isaacs, D., and Levinsky, R.J. (1982) *Clin. Exp. Im.* 50, 454–460.
75. Fischer, A., Descamps-Latscha, B., Gerota, I., Scheinmetzler, C., and Virelizier, J.L. (1983) *Lancet* 2, 473–476.
76. Harlan, J.M., Killen, P.D., Senecal, F.M., Schwartz, B.R., Yee, E.K., Taylor, R.F., Beatty, P.G., Price, T.H., and Ochs, H.D. (1985) *Blood* 66, 167–178.
77. Hayward, A.R., Leonard, J., Wood, C.B.S., Harvey, B.A.M., Greenwood, M.C., and Soothill, J.F. (1979) *Lancet* 1, 1099–1101.

78. Kobayashi, K., Fujita, K., and Okino, F. (1984) *Pediatrics* 73, 606–610.
79. Anderson, D.C. and Springer, T.A. (1987) *Ann. Rev. Med.* 38, 175–194.
80. Todd, III, R.F. and Freyer, D.R. (1988) *Hematol. Oncol. Clin. North Am.* 2, 13–28.
81. Springer, T.A., Thompson, W.S., Miller, L.J., Schmalstieg, F.C., and Anderson, D.C. (1984) *J. Exp. Med.* 160, 1901–1918.
82. Schmalstieg, F.C., Rudloff, H.E., Hillman, G.R., and Anderson, D.C. (1986) *J. Leuk. Biol.* 40, 677–691.
83. Anderson, D.C., Miller, L.J., Schmalstieg, F.C., Rothlein, R., and Springer, T.A. (1986) *J. Immunol.* 137, 15–27.
84. Dana, N., Styrt, B., Griffin, J.D., Todd, III, R.F., Klempner, M.S., and Arnaout, M.A. (1986) *J. Immunol.* 137, 3259–3263.
85. Detmers, P.A., Wright, S.D., Olsen, E., Kimball, B., and Cohn, Z.A. (1987) *J. Cell Biol.* 105, 1137–1145.
86. Philips, M., Buyon, J.P., Winchester, R., Weissmann, G., and Abramson, S. (1988) *J. Clin. Invest.* 82, 495–501.
87. Bainton, D.F., Miller, L.J., Kishimoto, T.K., and Springer, T.A. (1987) *J. Exp. Med.* 166, 1641–1653.
88. Jones, D.H., Schmalstieg, F.C., Dempsey, K., Krater, S.S., Nannen, D.D., Smith, C.W., and Anderson, D.C. (1990) *Blood* 75, 488–498.
89. Jones, D.H., Schmalstieg, F.C., Hawkins, H.K., Burr, B.L., Rudloff, H.E., Krater, S.S., Smith, C.W., and Anderson, D.C. (1989) *Leukocyte Adhesion Molecules: Structure, Function, and Regulation* (Springer, T.A., Anderson, D.C., Rosenthal, A.S., and Rothlein, R., eds.) pp. 106–124, Springer-Verlag, New York.
90. O'Shea, J.J., Brown, E.J., Seligmann, B.E., Metcalf, J.A., Frank, M.M., and Gallin, J.I. (1985) *J. Immunol.* 134, 2580–2587.
91. Petty, H.R., Francis, J.W., Todd, III, R.F., Petrequin, P.R., and Boxer, L.A. (1987) *J. Cell. Physiol.* 133, 235–242.
92. Stevenson, K.B., Nauseef, W.M., and Clark, R.A. (1987) *J. Immunol.* 139, 3759–3763.
93. Todd, III, R.F., Arnaout, M.A., Rosin, R.E., Crowley, C.A., Peters, W.A., and Babior, B.M. (1984) *J. Clin. Invest.* 74, 1280–1290.
94. Lo, S.K., Van Seventer, G.A., Levin, S.M., and Wright, S.D. (1989) *J. Immunol.* 143, 3325–3329.
95. Francis, J.W., Todd, R.F., Boxer, L.A., and Petty, H.R. (1989) *J. Cell. Physiol.* 140, 519–523.
96. Ryan, G.B., Borysenko, J.Z., and Karnovsky, M.J. (1974) *J. Cell Biol.* 62, 351–365.
97. Smith, C.W. and Hollers, J.C. (1980) *J. Clin. Invest.* 65, 804–812.
98. Springer, T.A. and Anderson, D.C. (1986) *Biochemistry of Macrophages.* pp. 102–126, Ciba Foundation Symposium, Pittman.
99. Stossel, T.P. (1978) *Leukocyte Chemotaxis* (Gallin, J.I. and Quie, P.G., eds.) pp. 143–147, Raven Press, New York.
100. Buyon, J.P., Abramson, S.B., Philips, M.R., Slade, S.G., Ross, G.D., Weissmann, G., and Winchester, R.J. (1988) *J. Immunol.* 140, 3156–3160.
101. Buyon, J.P., Philips, M.R., Abramson, S.B., Slade, S.G., Weissmann, G., and Winchester, R. (1989) *Leukocyte Adhesion Molecules: Structure, Function, and Regulation* (Springer, T.A., Anderson, D.C., Rosenthal, A.S., and Rothlein, R., eds.) pp. 72–83, Springer-Verlag, New York.

102. Lopez, A.F., Williamson, D.J., Gamble, J.R., Begley, C.G., Harlan, J.M., Klebanoff, S.J., Waltersdorph, A., Wong, G., Clark, S.C., and Vadas, M.A. (1986) *J. Clin. Invest.* 78, 1220–1228.
103. Schleiffenbaum, B., Moser, R., Patarroyo, M., and Fehr, J. (1989) *J. Immunol.* 142, 3537–3542.
104. Schwartz, B.R. and Harlan, J.M. (1989) *Biochem. Biophy. Res. Comm.* 165, 51–57.
105. Vedder, N.B. and Harlan, J.M. (1988) *J. Clin. Invest.* 81, 676–682.
106. Dransfield, I. and Hogg, N. (1989) *EMBO J.* 8, 3759–3765.
107. Furie, M.B., Tancinco, M.C.A., and Smith, C.W. (1991) *Blood* 78, 2089–2097.
108. de Fougerolles, A.R., Stacker, S.A., Schwarting, R., and Springer, T.A. (1991) *J. Exp. Med.* 174, 253–267.
109. Dustin, M.L., Rothlein, R., Bhan, A.K., Dinarello, C.A., and Springer, T.A. (1986) *J. Immunol.* 137, 245–254.
110. Munro, J.M., Pober, J.S., and Cotran, R.S. (1989) *Am. J. Pathol.* 135, 121–133.
111. Munro, J.M., Pober, J.S., and Cotran, R.S. (1991) *Lab. Invest.* 64, 295–299.
112. Pober, J.S., Gimbrone, Jr., M.A., Lapierre, L.A., Mendrick, D.L., Fiers, W., Rothlein, R., and Springer, T.A. (1986) *J. Immunol.* 137, 1893–1896.
113. Staunton, D.E., Dustin, M.L., and Springer, T.A. (1989) *Nature* 339, 61–64.
114. Diamond, M.S., Staunton, D.E., deFougerolles, A.R., Stacker, S.A., Garcia-Aguilar, J., Hibbs, M.L., and Springer, T.A. (1990) *J. Cell Biol.* 111, 3129–3139.
115. Kishimoto, T.K., Larson, R.S., Corbi, A.L., Dustin, M.L., Staunton, D.E., and Springer, T.A. (1989) *Adv. Immunol.* 46, 149–182.
116. Dustin, M.L. and Springer, T.A. (1989) *Nature* 341, 619–624.
117. Mentzer, S.J., Gromkowski, S.H., Krensky, A.M., Burakoff, S.J., and Martz, E. (1985) *J. Immunol.* 135, 9–11.
118. Patarroyo, M., Beatty, P.G., Fabre, J.W., and Gahmberg, C.G. (1985) *Scand. J. Immunol.* 22, 171–182.
119. Rothlein, R., Dustin, M.L., Marlin, S.D., and Springer, T.A. (1986) *J. Immunol.* 137, 1270–1274.
120. Rothlein, R. and Springer, T.A. (1986) *J. Exp. Med.* 163, 1132–1149.
121. Springer, T.A. (1990) *Nature* 346, 425–434.
122. Van Kooyk, Y., Van De Wiel-van Kemenade, P., Weder, P., Kuijpers, T.W., and Figdor, C.G. (1989) *Nature* 342, 811–813.
123. Van Kooyk, Y., Weder, P., Hogervorst, F., Verhoeven, A.J., Van Seventer, G.A., te Velde, A.A., Borst, J., Keizer, G.D., and Figdor, C.G. (1991) *J. Cell Biol.* 112, 345–354.
124. Kavanaugh, A.F., Lightfoot, E., Lipsky, P.E., and Oppenheimer-Marks, N. (1991) *J. Immunol.* 146, 4149–4156.
125. Oppenheimer-Marks, N., Davis, L.S., Bogue, D.T., Ramberg, J., and Lipsky, P.E. (1991) *J. Immunol.* In press.
126. Altieri, D.C. and Edgington, T.S. (1988) *J. Biol. Chem.* 263, 7007–7015.
127. Relman, D., Tuomanen, E., Falkow, S., Golenbock, D.T., Saukkonen, K., and Wright, S.D. (1990) *Cell* 61, 1375–1382.
128. Ross, G.D., Thompson, R.A., Walport, M.J., Springer, T.A., Watson, J.V., Ward, R.H.R., Lida, J., Newman, S.L., Harrison, R.A., and Lachmann, P.J. (1985) *Blood* 66, 882–890.

129. Russell, D.G. and Wright, S.D. (1988) *J. Exp. Med.* 168, 279–292.
130. Shappell, S.B., Toman, C., Anderson, D.C., Taylor, A.A., Entman, M.L., and Smith, C.W. (1990) *J. Immunol.* 144, 2702–2711.
131. Wright, S.D. and Jong, M.T.C. (1986) *J. Exp. Med.* 164, 1876–1888.
132. Wright, S.D., Lo, S.K., and Detmers, P.A. (1989) *Leukocyte Adhesion Molecules: Structure, Function and Regulation* (Springer, T.A., Anderson, D.C., Rothlein, R., and Rosenthal, A.S., eds.) pp. 190–207, Springer-Verlag, New York.
133. Wright, S.D., Rao, P.E., VanVoorhis, W.C., Craigmyle, L.S., Iida, K., Talle, M.A., Westbery, E.F., Goldstein, G., and Silverstein, S.C. (1983) *Proc. Nat. Acad. Sci. USA* 80, 5699–5703.
134. Cybulsky, M.I., McComb, D.J., and Movat, H.Z. (1989) *Am. J. Pathol.* 135, 227–235.
135. Allison, F., Smith, M.R., and Wood, W.B. (1955) *J. Exp. Med.* 102, 655–668.
136. Lawrence, M.B., Smith, C.W., Eskin, S.G., and McIntire, L.V. (1990) *Blood* 75, 227–237.
137. Worthen, G.S., Smedly, L.A., Tonnesen, M.G., Ellis, D., Voelkel, N.F., Reeves, J.T., and Henson, P.M. (1987) *J. Appl. Physiol.* 63, 2031–2041.
138. Anderson, D.C., Abbassi, O., Kishimoto, T.K., Koenig, J.M., McIntire, L.V., and Smith, C.W. (1991) *J. Immunol.* 146, 3372–3379.
139. Anderson, D.C., Rothlein, R., Marlin, S.D., Krater, S.S., and Smith, C.W. (1990) *Blood* 78, 2613–2621.
140. Jutila, M.A., Rott, L., Berg, E.L., and Butcher, E.C. (1989) *J. Immunol.* 143, 3318–3324.
140a. Kishimoto, T.K., Jutila, M.A., Berg, E.L., and Butcher, E.C. (1989) *Science* 245, 1238–1241.
141. Stacker, S.A. and Springer, T.A. (1991) *J. Immunol.* 146, 648–655.
142. Arnaout, M.A., Lanier, L.L., and Faller, D.V. (1988) *J. Cell. Physiol.* 137, 305–313.
143. Beekhuizen, H., Tilburg, J.C., and VanFurth, R. (1990) *J. Immunol.* 145, 510–518.
144. Keizer, G.D., te Velde, A.A., Schwarting, R., Figdor, C.G., and De Vries, J.E. (1987) *Eur. J. Immunol.* 17, 1317–1322.
145. Luscinskas, F.W., Brock, A.F., Arnaout, M.A., and Gimbrone, Jr., M.A. (1989) *J. Immunol.* 142, 2257–2263.
146. Bevilacqua, M.P., Pober, J.S., Mendrick, D.L., Cotran, R.S., and Gimbrone, Jr., M.A. (1987) *Proc. Natl. Acad. Sci. USA* 84, 9238–9242.
147. Bevilacqua, M.P., Stengelin, S., Gimbrone, Jr., M.A., and Seed, B. (1989) *Science* 243, 1160–1165.
148. Picker, L.J., Kishimoto, T.K., Smith, C.W., Warnock, R.A., and Butcher, E.C. (1990) *Nature* 349, 796–799.
149. Diamond, M.S., Staunton, D.E., Marlin, S.D., and Springer, T.A. (1991) *Cell* 65, 961–971.
150. Clark, E.A., Ledbetter, J.A., Holly, R.C., Dinndorf, P.A., and Shu, G. (1986) *Human Immunol.* 16, 100–113.
151. Staunton, D.E., Dustin, M.L., Erickson, H.P., and Springer, T.A. (1990) *Cell* 61, 243–254.
152. Entman, M.L., Youker, K., Shoji, T., Taylor, A.A., Shappell, S.B., and Smith, C.W. (1991) *Clin. Res.* 39, 159A.

153. Nathan, C.F. (1987) *J. Clin. Invest.* 80, 1550–1560.
154. Nathan, C.F., Srimal, S., Farber, C., Sanchez, E., Kabbash, L., Asch, A., Gailit, J., and Wright, S.D. (1989) *J.Cell Biol.* 109, 1341–1349.
155. Hawkins, H.K., Heffelfinger, S., and Anderson, D.C. (1991) *Pediatr. Pathol.* In press.
156. Doerschuk, C.M., Winn, R.K., Coxson, H.O., and Harlan, J.M. (1990) *J. Immunol.* 144, 2327–2333.
157. Furie, M.B., Burns, M.J., Tancinco, M.C.A., Benjamin, C.D., and Lobb, R.R. (1991) *J. Immunol.* In press.
158. Benjamin, C., Dougas, I., Chi-Rosso, G., Luhowskyj, S., Rosa, M., Newman, B., Osborn, L., Vassallo, C., Hession, C., Goelz, S., McCarthy, K., and Lobb, R. (1990) *Biochem. Biophys. Res. Comm.* 171, 348–353.
159. Lo, S.K., Lee, S., Ramos, R.A., Lobb, R., Rosa, M., Chi-Rosso, G., and Wright, S.D. (1991) *J. Exp. Med.* 173, 1493–1500.
160. Atherton, A. and Born, G.V.R. (1972) *J. Physiol.* 222, 447–474.
161. Atherton, A. and Born, G.V.R. (1973) *J. Physiol.* 233, 157–165.
162. Fiebig, E., Ley, K., and Arfors, K.E. (1991) *Int. J. Microcirc. Clin. Exp.* 10, 127–144.
163. Ley, K., Cerrito, M., and Arfors, K.E. (1991) *Am. J. Physiol.* 260, H1667–H1673.
164. Ley, K., Lundgren, E., Berger, E., and Arfors, K.E. (1989) *Blood* 73, 1324–1330.
165. Hernandez, L.A., Grisham, M.B., Twohig, B., Arfors, K.E., Harlan, J.M., and Granger, D.N. (1987) *Am. J. Physiol.* 238, H699–H703.
166. House, S.D. and Lipowsky, H.H. (1987) *Microvasc. Res.* 34, 363–379.
167. Zimmerman, B.J. and Granger, D.N. (1990) *Am. J. Physiol.* 259, H390–H394.
168. Arfors, K.E., Lundberg, C., Lindbom, L., Lundberg, K., Beatty, P.G., and Harlan, J.M. (1987) *Blood* 69, 338–340.
169. Smith, C.W., Marlin, S.D., Rothlein, R., Lawrence, M.B., McIntire, L.V., and Anderson, D.C. (1989) *Leukocyte Adhesion Molecules: Structure, Function, and Regulation* (Springer, T.A., Anderson, D.C., Rosenthal, A.S., and Rothlein, R., eds.) pp. 170–189, Springer-Verlag, New York.
170. Perry, M.A. and Granger, D.N. (1991) *J. Clin. Invest.* 87, 1798–1804.
171. Kishimoto, T.K., Jutila, M.A., and Butcher, E.C. (1990) *Proc. Natl. Acad. Sci. USA* 87, 2244–2248.
172. Spertini, O., Kansas, G.S., Reimann, K.A., Mackay, C.R., and Tedder, T.F. (1991) *J. Immunol.* 147, 942–949.
173. Yednock, T.A. and Rosen, S.D. (1989) *Adv. Immunol.* 44, 313–378.
174. Picker, L.J., Warnock, R.A., Burns, A.R., Doerschuk, C.M., Berg, E.L., and Butcher, E.C. (1991) *Cell* 66, 921–933.
175. von Andrian, U.H., Chambers, J.D., McEvoy, L.M., Bargatze, R.F., Arfors, K.E., and Butcher, E.C. (1991) *Proc. Natl. Acad. Sci. USA* 88, 7538–7542.
176. Ley, K., Gaehtgens, P., Fennie, C., Singer, M.S., Lasky, L.A., and Rosen, S.D. (1991) *Blood* 77, 2553–2555.
177. Lewinsohn, D.M., Bargatze, R.F., and Butcher, E.C. (1987) *J. Immunol.* 138, 4313–4321.
178. Watson, S.R., Fennie, C., and Lasky, L.A. (1991) *Nature* 349, 164–167.

179. Hallmann, R., Jutila, M.A., Smith, C.W., Anderson, D.C., Kishimoto, T.K., and Butcher, E.C. (1991) *Biochem. Biophys. Res. Commun.* 174, 236–243.
180. Jutila, M.A., Kishimoto, T.K., and Butcher, E.C. (1990) *Blood* 75, 1–5.
181. Mulligan, M.S., Varani, J., Dame, M.K., Lane, C.L., Smith, C.W., Anderson, D.C., and Ward, P.A. (1991) *J. Clin. Invest.* 88, 1396–1406.
182. Gundel, R.H., Wegner, C.D., Torcellini, C.A., Clarke, C.C., Haynes, N., Rothlein, R., Smith, C.W., and Letts, L.G. (1991) *J. Clin. Invest.* 88, 1407–1411.
183. Dobrina, A., Carlos, T.M., Schwartz, B.R., Beatty, P.G., Ochs, H.D., and Harlan, J.M. (1990) *Immunol.* 69, 429–438.
184. Lowe, J.B., Stoolman, L.M., Nair, R.P., Larsen, R.D., Berhend, T.L., and Marks, R.M. (1990) *Cell* 63, 475–484.
185. Phillips, M.L., Nudelman, E., Gaeta, F.C.A., Perez, M., Singhal, A.K., Hakomori, S., and Paulson, J.C. (1990) *Science* 250, 1130–1132.
186. Walz, G., Aruffo, A., Kolanus, W., Bevilacqua, M.P., and Seed, B. (1990) *Science* 250, 1132–1135.
187. McEver, R.P., Beckstead, J.H., Moore, K.L., Marshall-Carlson, L., and Bainton, D.F. (1989) *J. Clin. Invest.* 84, 92–99.
188. Bonfanti, R., Furie, B.C., Furie, B., and Wagner, D.D. (1989) *Blood* 73, 1109–1112.
189. Hattori, R., Hamilton, K.K., Fugates, R.D., McEver, R.P., and Sims, P.J. (1989) *J. Biol. Chem.* 264, 7768–7771.
190. Geng, J.G., Bevilacqua, M.P., Moore, K.L., McIntyre, T.M., Prescott, S.M., Kim, J.M., Bliss, G.A., Zimmerman, G.A., and McEver, R.P. (1990) *Nature* 343, 757–760.
191. Patel, K.D., Zimmerman, G.A., Prescott, S.M., McEver, R.P., and McIntyre, T.M. (1991) *J. Cell Biol.* 112, 749–759.
192. Lorant, D.E., Patel, K.D., McIntyre, T.M., McEver, R.P., Prescott, S.M., and Zimmerman, G.A. (1991) *J. Cell Biol.* 115, 223–234.
193. Lawrence, M.B. and Springer, T.A. (1991) *Cell* 65, 1–20.
194. Worthen, G.S., Schwab III, B., Elson, E.L., and Downey, G.P. (1989) *Science* 245, 183–186.
195. Oppenheimer-Marks, N. and Ziff, M. (1988) *Cell. Immunol.* 114, 307–323.
196. Van Epps, D.E., Potter, J., Vachula, M., Smith, C.W., and Anderson, D.C. (1989) *J. Immunol.* 143, 3207–3210.
197. Shimizu, Y., Newman, W., Gopal, T.V., Horgan, K.J., Graber, N., Beall, L.D., Van Seventer, G.A., and Shaw, S. (1991) *J. Cell Biol.* 113, 1203–1212.
198. Keizer, G.D., Visser, W., Vliem, M., and Figdor, C.G. (1988) *J. Immunol.* 140, 1393–1400.
199. Graber, N., Gopal, T.V., Wilson, D., Beall, L.D., Polte, T., and Newman, W. (1990) *J. Immunol.* 145, 819–830.
200. Shimizu, Y., Shaw, S., Graber, N., Gopal, T.V., Horgan, K.J., Van Seventer, G.A., and Newman, W. (1991) *Nature* 349, 799–802.
201. Larsen, C.G., Anderson, A.O., Appella, E., Oppenheim, J.J., and Matsushima, K. (1989) *Science* 243, 1464–1466.
202. Spertini, O., Kansas, G.S., Munro, J.M., Griffin, J.D., and Tedder, T.F. (1991) *Nature* 349, 691–694.
203. Shimizu, Y., Van Seventer, G.A., Horgan, K.J., and Shaw, S. (1990) *Nature* 345, 250–253.

204. Bochner, B.S., Luscinskas, F.W., Gimbrone, Jr., M.A., Newman, W., Sterbinsky, S.A., Derse-Anthony, C.P., Klunk, D., and Schleimer, R.P. (1991) *J. Exp. Med.* 173, 1553–1556.
205. Lamas, A.M., Mulroney, C.M., and Schleimer, R.P. (1988) *J. Immunol.* 140, 1500–1505.
206. Wegner, C.D., Gundel, R.H., Reilly, P., Haynes, N., Letts, L.G., and Rothlein, R. (1990) *Science* 247, 456–459.
207. Kyan-Aung, U., Haskard, D.O., Poston, R.N., Thornhill, M.H., and Lee, T.H. (1991) *J. Immunol.* 146, 521–528.
208. Walsh, G.M., Mermod, J.J., Hartnell, A., Kay, A.B., and Wardlaw, A.J. (1991) *J. Immunol.* 146, 3419–3423.

CHAPTER 6

In Vivo Models of Leukocyte Adherence to Endothelium

John M. Harlan, Robert K. Winn, Nicholas B. Vedder, Claire M. Doerschuk, Charles L. Rice

The study of leukocyte–endothelial interactions, particularly the cell surface adhesion molecules involved, has become one of the most active and productive areas in cell biology. Ten years ago, only a single adhesion protein had been identified immunologically—the murine lymphocyte peripheral lymph node homing receptor identified by the monoclonal antibody (MAb) MEL-14.[1] Presently, over a dozen leukocyte and endothelial cell adhesion proteins have been immunochemically characterized and molecularly cloned.[2] They comprise two categories of adhesion molecules: leukocyte integrins that bind to ligands on the endothelium that are members of the immunoglobulin superfamily (see Chapter 1), and selectins on both leukocytes and endothelial cells that recognize specific carbohydrate counter-structures (see Chapter 2). The function of these proteins in leukocyte binding to endothelium has largely been defined by studies examining binding of isolated leukocytes to frozen sections of vessels in tissue preparations or to cultured endothelium. These *ex vivo* and *in vitro* studies have been invaluable in identifying candidate adhesion proteins, elucidating factors controlling their regulation, and suggesting potential cellular interactions. However, in the absence of a genetic deficiency such as the CD11/CD18-deficiency syndrome,[3] the relative importance of a particular adhesion pathway in leukocyte adherence to endothelium and emigration in inflammatory and immune reactions cannot be established until *in vivo* studies are performed. Monoclonal antibodies that bind to functional epitopes of adhesion proteins and inhibit leukocyte binding to endothelium have proven to be critical reagents in this regard. This chapter will review various *in vivo* models that have examined leukocyte or endothelial cell adhesion proteins, particularly those that have utilized blocking MAbs. It will focus on leukocyte adhesion to endothelium in inflammation and immune reactions rather than the physiologic homing of lymphocytes to lymphoid tissue.[4]

INTRAVITAL MICROSCOPY

Leukocyte Sticking

Elegant *in vivo* observations by a number of investigators have established white cell "sticking" to endothelium as a hallmark of the acute inflammatory response. According to Grant,[5] Dutrochet in 1824 was the first investigator to describe the "sticking" and emigration of white cells, which he described as "vesicular globules."[6]

> "... the vesicular globules contained in the blood are added to the tissues of the organs and become fixed there to augment and repair them so that nutrition consists of a veritable intercalation of fully formed and extremely tiny cells. ... *Many times I have seen blood cells leaving the blood stream, being arrested and becoming fixed to the organic tissue.* I have seen the phenomenon, which I was far from suspecting, when I observed the movement of the blood in the transparent tail of young tadpoles under the microscope... Observing the movement of the blood, I have seen many times a single cell escape laterally from the blood vessel and move in the transparent tissue ... with a slowness which contrasted strongly with the rapidity of the circulation from which the cells had escaped. Soon afterwards, the cell stopped moving and remained fixed in the transparent tissue. A comparison with the granulations which this tissue contained showed that they were in no way different. There is no doubt that these semi-transparent granulations were also blood cells which had previously become fixed." [From Grant;[5] italics added]

As Movat notes,[7] Cohnheim (1889) elegantly described the phenomenon of leukocyte sticking to endothelium based upon intravital microscopy of inflammatory reactions in the frog mesentery or tongue:[8]

> "This stage having been reached, the vessels are seen to be all of them very wide; a multitude of capillaries which were formerly hardly perceptible can now be clearly distinguished... *But it is the veins rather than the capillaries that attract the notice of the observer*, for slowly and gradually there is developed in them an extremely characteristic condition; the originally plasmatic zone becomes filled

with innumerable colorless corpuscles. The plasmatic zone of the veins, you will remember, is always occupied by scattered colorless blood-corpuscles, which owing to their globular form and low specific gravity are driven into the periphery of the stream, and *whose adhesiveness makes it difficult for them to escape from the wall once they have come into contact with it.* It is obvious that this difficulty will be enhanced in proportion to the slowness of the blood-stream; and thus it is not surprising that a gradual accumulation of large numbers of colorless corpuscles should take place in the peripheral zone, and here come to be comparatively motionless . . . Yet this does not lessen the striking contrast presented by the central column of red blood corpuscles, flowing on in an uninterrupted stream of uniform velocity . . . the *internal surface of the vein appears paved with a single but unbroken layer of colorless corpuscles* without the interposition at any time of a single red one. It is the separation of the white from the red corpuscles that gives the venous stream in these cases that characteristic appearance, anything analogous to which you will look for in vain in the other vessels . . ." [From Movat;[7] italics added]

It was a matter of controversy whether leukocyte sticking was mediated by adherence changes in the endothelium or the leukocyte. As discussed by Grant[5] and Movat,[7] Cohnheim (1889) emphasized molecular alterations in the vessel wall as a major cause of leukocyte sticking and considered subsequent emigration of leukocytes to be a process of mechanical filtration rather than active movement.[8] In contrast, Metchnikoff (1893) believed ". . . that the migration is effected by the amoeboid power of the leukocytes,"[9] citing the studies of Leber (1888) who was the first to use the term *chemotaxis* for the directed movement of leukocytes.[5] Perhaps not surprisingly, it now appears that both investigators were correct as *in vitro* and *in vivo* studies show that there are both endothelial- and leukocyte-dependent mechanisms of adhesion.

The temporal sequence of events occurring during leukocyte adherence and emigration in acute inflammation was carefully described by Clark and Clark[10] and subsequently by Allison and colleagues[11] in studies utilizing the Sandison transparent ear chamber in rabbits. As discussed by Movat,[7] Clark and Clark detailed the effect of a globule of croton oil deposited in the chamber:[10]

"(A leukocyte) could be seen to glide smoothly along or to *roll over and over along the wall, depending upon the speed of the circulation at the time, until it arrived at a particular point, when it stopped and remained adherent to the wall for a few seconds or several minutes.* After being detached by the force of the stream in the manner just described, it resumed its former rounded or oval shape and continued to glide or roll along smoothly as before. This was repeated in the case of each leukocyte which came in contact with the vessel wall at this particular point . . . , but *once the leukocytes had passed the focus of stickiness they no longer adhered to the endothelium.*" [italics added]

Allison et al.[11] detailed leukocyte–vessel wall interactions following a thermal injury:

"Three points in regard to the sticking of leukocytes deserve particular emphasis. First, vasodilation does not always precede the sticking of leukocytes since white cell adherence was observed to occur during vasoconstriction . . . Secondly, during the course of the inflammatory reaction *leukocytes were frequently seen to stick to one another, indicating that the increased adhesiveness characteristic of the inflammatory response is not limited to the endothelium* . . . Thirdly, it was repeatedly noted that leukocytes dislodged into the circulation from localized areas of sticky endothelium did not adhere to the walls of the undamaged vessels into which they escaped. *Thus leukocytes, although capable of becoming sticky, do not attach themselves to the surfaces of uninjured endothelium.* [italics added]

The molecular basis of these adhesive interactions of leukocytes and endothelium *in vivo* was not established until the past decade. There is now compelling evidence that the CD11/CD18 complex is necessary for leukocyte sticking at sites of inflammation *in vivo*. This was first shown by Arfors et al.,[12] who examined leukocyte sticking to microvascular endothelium in the rabbit tenuissimus muscle by intravital microscopy. Application of leukotriene B_4 (LTB_4) or zymosan-activated serum induced vigorous leukocyte adherence to the endothelium of postcapillary venules and subsequent extravasion into the interstitial space. Pretreatment of animals with a single injection of the CD18 MAb 60.3 developed by Beatty et al.[13] completely inhibited leukocyte sticking and emigration in response to application of LTB_4 or zymosan-activated serum (Figure 6-1). Circulating leukocyte counts were not affected by MAb treatment, and $F(ab)_2$ fragments of MAb were active. These results were subsequently confirmed by Lindbom et al.,[14] Argenbright et al.,[15] and Perry and Granger[16] using different CD18 MAbs. Argenbright et al. found that a CD11a (anti-LFA-1α) MAb and a CD54 (anti-ICAM-1) MAb also strongly inhibited stimulated adherence, while a CD11b (anti-Mac-1α) MAb produced significant but weaker inhibition. Interestingly, both the CD18 MAb and the CD11a MAb were found to displace adherent leukocytes, whereas the CD11b and CD54 MAbs did not. These results suggest that the CD11a and CD11b subunits of the CD11/CD18 complex and the endothelial ligand ICAM-1 are important in leukocyte sticking induced by chemoattractants.

Perry and Granger[16] examined the role of CD11/CD18 in leukocyte adherence at various shear rates in the cat mesentery. Previous *in vitro* studies by several investigators had indicated that activated leukocytes did not adhere to cultured endothelial monolayers at shear rates $\geq 2-3$ dynes/cm^2.[17-19] In contrast, Perry and Granger found that CD11/CD18-dependent adherence (i.e., adherence inhibited by a CD18 MAb) was present at shear rates up to 15 dynes/cm^2 (with the normal range of shear stress in the cat mesentery venules estimated between 3 and 36 dynes/cm^2). They suggested that the discrepancy between the *in vivo* results and those obtained *in vitro* may result from one or more of the following factors: 1) the medium used *in vitro* does not adequately mimic *in vivo* conditions, 2) cultured endothelial cells may not express the ligand for CD11/CD18 at sufficient density, 3) *in vitro* studies

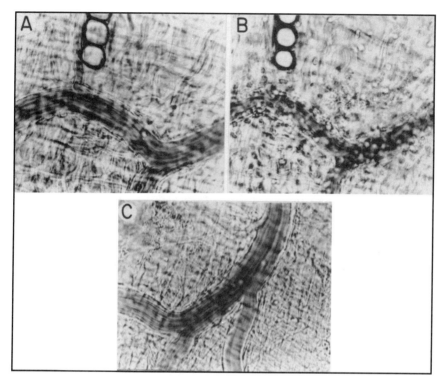

Figure 6-1. Intravital microscopy of rabbit hindlimb microvasculature. *A* shows a postcapillary venule in the rabbit tenuissimus muscle of a control animal. *B* shows a venule in a control animal following topical application of zymosan-activated serum. *C* shows a venule in an animal pretreated with the CD18 MAb 60.3 (2 mg/kg) after a 20-minute topical application of zymosan-activated serum. In contrast to *B*, there are no adherent or extravasated leukocytes in *C*. Photomicrographs courtesy of Karl E. Arfors, La Jolla, CA. Adapted from Arfors, K.E., Lundberg, D., Lindbom, L., Lundberg, K., Beatty, P.G., and Harlan, J.M. (1987) *Blood* 69, 338–340, with permission.

employ large vessel passaged endothelium from umbilical vein or arterial endothelium rather than venular endothelium, or 4) isolation of leukocytes for *in vitro* studies may alter their adhesive properties. An additional possibility is that the experimental conditions employed in the *in vitro* studies did not recruit the LECAM-1(LAM-1) or CD62 (GMP-140, PADGEM) selectin molecules that may contribute to the initial slowing or arrest of leukocytes *in vivo*, allowing subsequent engagement of CD11/CD18 and shear stress-resistant sticking.[20]

The studies of Perry and Granger[16] also addressed the role of CD11/CD18 in leukocyte sticking to arterioles versus venules. Virtually all investigators using intravital microscopy have observed that leukocytes adhere preferentially to venules rather than to arterioles. The higher shear rate found in arterioles has been proposed as an explanation for reduced leukocyte adherence. However, when shear rate was controlled experimentally, Perry and Granger found that leukocyte adherence to venules was much greater than to arterioles at any given shear rate. The low level of leukocyte adherence to arterial endothelium at low shear rates was inhibited by CD18 MAb, leading the researchers to suggest that the difference between venules and arterioles may be a lower density of ligand for CD11/CD18 on arteriolar endothelium.[16]

Leukocyte Rolling

Intravital microscopy of surgically-prepared tissue reveals two types of leukocyte adherence to venular endothelium: firm adhesion or sticking and transient adhesion or rolling.[21] Although often noted by earlier investigators, the phenomenon of leukocyte rolling was first carefully studied by Atherton and Born in the early 1970s.[22, 23] They developed a method for quantifying the interaction of and circulating leukocytes and endothelium by counting the rolling cells in small venules in the hamster cheek pouch or mouse mesentery under various experimental conditions. They found that:

- The rolling of leukocytes increased transiently after the preparation was first set up and then fell to a constant low value.
- Rolling was abolished by superfusion with ethylendiamine tetra-acetate (EDTA), suggesting that it was dependent upon calcium or magnesium ions.
- The mean velocity of rolling leukocytes increased in proportion to mean blood flow velocity between blood velocities of about 300 and 1000 µm/s, whereas above this the velocity of the cells did not increase further.

The researchers proposed the following explanation for rolling:[23]

> "As the simplest explanation of granulocytes rolling, it has been proposed (Atherton and Born, 1972) that circulating granulocytes collide randomly with venule walls and that these collisions behave elastically when the venule is normal but inelastically when it is inflamed. *As a result of biochemical changes effected in the walls by inflammatory agents, granulocytes colliding with the walls experience an adhesive force as well as the shear force exerted at the wall by the blood flow. The rolling movement is then the resultant of these two forces.*" [italics added]

Leukocyte rolling has been reported to be absent when vessels are observed without surgical manipulation as in the intact bat wing.[24] Fiebig et al.,[21] however, noted that rolling was rapidly induced upon surgical preparation and tissue handling, and suggested that "spontaneous" rolling in surgically-prepared tissue represents an early response to mild inflammation.

Intravital microscopy demonstrates that leukocyte rolling precedes adherence at sites of inflammation. Perry and Granger investigated the relationship between leukocyte rolling velocity and leukocyte adherence.[16] They found that adherence was minimal at leukocyte rolling velocities above 50 µm/s, but increased dramatically once the velocity fell below 50 µm/s, with maximal adherence at 5–20 µm/s. They suggested that the progressive decrease in rolling velocity leads to a progressive increase in contact time between adhesion molecules expressed on the surface of the leukocyte or endothelial cell. With sufficient time of contact, firm adhesion occurs.

The role of CD11/CD18 complex in leukocyte rolling has been examined by intravital microscopy. Arfors et al.[12] first reported that pretreatment of rabbits with a CD18 MAb had no effect on leukocyte rolling induced by application of a chemoattractant to the local venule. Similar results were observed by Lindbom et al.[14] Indeed, Perry and Granger[16] found that pretreatment with a CD18 MAb significantly increased leukocyte rolling velocity. Thus, although CD11/CD18 is necessary for leukocyte sticking, it does not appear to be involved in rolling.

Several observations have indirectly suggested a role for selectin receptors in leukocyte rolling. Rolling is cation-dependent,[22] sensitive to neuraminadase,[25] and inhibited by sulfated polysaccharides[26]—all characteristics of selectin-mediated adhesive interactions. Ley et al.[27] examined the role of the L-selectin (LECAM-1, LAM-1) in leukocyte rolling in the venules of rat mesentery. They utilized a polyclonal antiserum to LECAM-1 and a recombinant soluble chimera of the murine homing receptor LECAM-1 and IgG (LEC-IgG). Microinfusion of the anti-LECAM-1 antiserum or LEC-IgG into venules markedly reduced the number of rolling leukocytes, whereas pre-immune serum or CD14-IgG control reagents were without effect. These investigators proposed that the inhibitory effect of anti-LECAM-1 antiserum and LEC-IgG on leukocyte rolling results from an inhibition of this selectin's interaction with an as yet unidentified, constitutively-expressed, or rapidly-induced endothelial ligand.

Using intravital microscopy of rabbit mesenteric venules, von Andrian et al.[28] reached a similar conclusion regarding the role of LECAM-1 in leukocyte rolling. They found that the anti-LECAM-1 MAb or its monovalent Fab fragments inhibited leukocyte rolling along the vessel wall. The researchers also proposed that reversible rolling mediated by LECAM-1 is an initial event that is important in leukocyte recruitment.

In view of these exciting results with LECAM-1-directed reagents, it will be of great interest to determine whether MAbs to the P-selectin (CD62, GMP-140, PADGEM), a rapidly-mobilized endothelial selectin, also inhibit leukocyte rolling along venules *in vivo*.

The Marginated Pool and Margination

There is a dynamic equilibrium between neutrophils within the circulation and those in the marginated pool. The marginated cells are within the vascular compartment at a concentration greater than that in the circulating blood. The marginal pool has been estimated to contain 0.6–3 times the total number of circulating neutrophils.[29] A marginated pool exists for monocytes[30-32] and lymphocytes[32,33] as well as neutrophils. The microvasculature of the lung, specifically the capillary bed, has been established as a major site of the marginated pool for neutrophils.[29,34,35] Retention of neutrophils and lymphocytes in the pulmonary microvasculature has been shown to increase when blood flow is reduced, indicating that pulmonary flow has a marked effect upon the uptake and release of leukocytes from the marginated pool.[36] Leukocyte size and deformability are also important factors in determining the size of the marginated pool because they affect the ability of leukocytes to transit the pulmonary microvasculature.[37] The mean diameter of the circulating leukocyte is 6–8 μm,[38] whereas that of the pulmonary capillaries is only 4–7 μm.[39,40] Leukocytes must deform, therefore, in order to transit through the pulmonary microcirculation.[37] Although of comparable size (7.5 μm), red cells are many times more deformable than leukocytes.[41] Consequently, it takes a leukocyte far longer than a red cell to transit the same pulmonary capillary segment. The multisegmented nature of the pulmonary microvasculature allows the more deformable red cell to stream past the less deformable neutrophil that is temporarily obstructing a capillary segment.[40] This results in a concentration of neutrophils with respect to red cells within the pulmonary microvasculature to produce the marginated pool[29]—without invoking any adhesive phenomena such as rolling or sticking.

Leukocytes in the marginal pool can be rapidly mobilized during periods of stress. Injection of epinephrine in normal individuals produces a marked increase in the circulating leukocyte count.[42] Two patients with CD11/CD18-deficiency were reported to exhibit a normal increment in circulating leukocyte count following injection of epinephrine,[43,44] suggesting that these patients had normal marginated pools and that the CD11/CD18 complex is not involved in this process. The fact that administration of a CD18 MAb to normal animals does not produce an increase in the circulating leukocyte count[45] and that CD11/CD18-deficient dog neutrophils marginate similarly to normal cells[46] also argues against a role for CD11/CD18 complex in this physiologic margination.

Intravenous infusion of chemotactic agents such as bacterial chemotactic peptide, zymosan-activated serum, or endotoxin that directly stimulates neutrophils produces a rapid, profound decrease in the circulating neutrophil count within five minutes with a return to normal after several hours. This acute neutropenia or "margination" results from sequestration of neutrophils within the pulmonary microvasculature. The stimuli that provoke margination also increase the adhesivity of neutrophils for endothelium via CD11/CD18 *in vitro*, raising the possibility that the CD11/CD18 complex mediates margination by increasing neutrophil adherence to pulmonary microvascular endothelium. This question was examined by Lundberg and Wright[47] and by

Doerschuk et al.[48] Both groups found that pretreatment of rabbits with a CD18 MAb did not prevent the immediate fall in neutrophil count following infusion of zymosan-activated plasma or bacterial chemotactic peptide. Moreover, Yong et al.[49] reported that infusion of granulocyte-macrophage colony stimulating factor in a patient with partial CD11/CD18-deficiency produced a decrease in the circulating leukocyte count similar to that observed in normal individuals. As suggested by Doerschuk et al.[50] and Worthen et al.,[51] pulmonary sequestration of neutrophils, produced by infusion of neutrophil activating agents, could result from a further decrease in neutrophil deformability, causing an increased retention in pulmonary microvasculature. However, Yoder et al.[46] showed that infusion of zymosan-activated plasma increased the transit time through the lung of normal neutrophils but not CD11/CD18-deficient neutrophils. Also, Doerschuk et al.[48] found that, although the CD18 MAb did not prevent the initial fall in circulating neutrophils at four minutes following infusion of zymosan-activated plasma, it prevented the sustained neutropenia and the accumulation of neutrophils in the lungs after 15 minutes of infusion of zymosan-activated plasma. It thus appears that the onset of neutropenia is due to the transient sequestration of neutrophils in the pulmonary capillaries through a CD11/CD18-independent process, likely to involve a decrease in deformability.[50,51] This is followed by a transient increase in neutrophil adhesiveness mediated by CD11/CD18 that is required to keep the sequestered population within the pulmonary microvasculature.[46,48]

EMIGRATION

Emigration of leukocytes from the bloodstream to extravascular sites of inflammation or immune reaction involves a complex but coordinated sequence of events.[52] Signals are generated at the extravascular site of infection or tissue damage (e.g., bacterial endotoxin, activated complement fragments, or cytokines) that activate the leukocyte and/or the endothelial cell. As a consequence of activation, one or both cell types becomes adhesive, leading initially to transient adhesion (rolling) and later to firm adhesion (sticking). Adherent leukocytes then move across the endothelial cell surface, diapedese between endothelial cells, and migrate through the subendothelial matrix to the extravascular site of inflammation or immune reaction. Leukocyte adherence to endothelium is a pivotal event in this cascade because adherence is necessary for subsequent diapedesis and migration. Consequently, defects in leukocyte adherence to endothelium associated with a congenital deficiency of adhesion proteins such as CD11/CD18 or induced by MAbs that block adhesion proteins are manifested *in vivo* by reduced accumulation of leukocytes at extravascular sites.[53] When interpreting results of *in vivo* studies examining emigration, however, it is important to remember that disruption of the cascade at any step—signal generation, activation, adherence, transendothelial migration, or migration through connective tissue—will prevent leukocyte emigration to tissue. Thus, a MAb to a putative leukocyte or endothelial adhesion protein could potentially inhibit leukocyte emigration by inhibiting cell activation, transendothelial migration, or leukocyte binding to matrix tissue as well as by blocking adherence to endothelium. Also, it is important to establish that binding of the MAb does not induce leukopenia by activating cells, provoking agglutination or aggregation, fixing complement, or promoting clearance by Fc receptors, because MAb-induced leukopenia will also result in decreased emigration.

CD11/CD18

The CD11/CD18-deficiency syndrome is characterized by a profound defect in phagocyte emigration.[3] Despite a moderate, chronic, neutrophilic leukocytosis with striking elevations of neutrophils during episodes of acute infection, neutrophils and monocytes of severely affected patients do not accumulate at sites of infection. Biopsies of tissues from sites of infection or inflammation show an absence or paucity of neutrophils and monocytes[54–57] and neutrophils and monocytes do not migrate to skin windows.[56–58] *In vitro* studies in 1984–1985[59,60] first demonstrated that CD11/CD18-deficient neutrophils or normal neutrophils treated with a blocking CD18 MAb failed to adhere to or migrate across cultured endothelium when stimulated with a variety of agonists, suggesting that the profound defect in phagocyte emigration observed in CD11/CD18-deficient patients resulted from impaired adherence of neutrophils to endothelium at sites of inflammation.

The critical role of CD11/CD18 in neutrophil adherence to endothelium and emigration to tissue was first demonstrated *in vivo* using intravital microscopy by Arfors et al.[12] They showed that superfusion of vessels in the microcirculation of the rabbit hind limb with a chemoattractant induced leukocyte adherence to endothelium and subsequent emigration to tissue, and that pretreatment of the animal with a single injection of the CD18 MAb 60.3[13] completely inhibited these responses (see Figure 6-1). In the same study they demonstrated that the CD18 MAb prevented neutrophil accumulation in the skin following intradermal injection of chemoattractants. Price et al.[45] demonstrated that the same CD18 MAb prevented neutrophil accumulation in polyvinyl sponges implanted subcutaneously in rabbits. Subsequently, CD18 MAbs were demonstrated to inhibit neutrophil emigration in a variety of animal models.[14,15,61–70]

In vitro studies suggest that the CD11b/CD18 (Mac-1, Mo1) heterodimer is primarily involved in stimulated neutrophil binding to endothelium,[60] although the CD11a/CD18 (LFA-1) heterodimer also contributes.[71–73] Rosen and colleagues demonstrated that a CD11b MAb heterodimer inhibited recruitment of neutrophils and monocytes to inflammatory sites in the peritoneum,[74] lung,[75] and footpad of the mouse.[76]

Results obtained in the animal models with blocking MAbs to functional epitopes of CD11/CD18 have been largely predictable from the profound defect in phagocyte emigration observed in CD11/CD18-deficient patients. One exception, however, comes from the studies of Doerschuk and colleagues who examined neutrophil emigration into the rabbit lung in response to various inflammatory stimuli.[66] They found that the CD18 MAb 60.3 virtually abolished neutrophil emigration into the rabbit lung in response to phorbol ester or *Esherichia coli* endotoxin, but was without significant effect on neutrophil accumulation in the alveoli induced by *Streptococcus pneumoniae* organisms or hydrochloric acid. When *S pneumoniae* organisms were placed in one lung and *E coli* endotoxin or phorbol ester was placed in the opposite lung of the same animal, pretreatment with the CD18 MAb markedly reduced neutrophil accumulation in the lung receiving *E coli* endotoxin or phorbol ester, but was without significant effect on neutrophil accumulation in the opposite lung receiving *S pneumoniae*. In some animals, *S pneumoniae* organisms were instilled into the trachea and simultaneously placed in the abdominal wall. Strikingly, pretreatment with the CD18 MAb completely inhibited neutrophil accumulation in the abdominal wall, but was without effect on neutrophil emigration in the lung (Figure 6-2). These surprising results in the rabbit model suggested that: 1) the mechanism of neutrophil emigration in the lung is stimulus-specific, that is, CD18-dependent for phorbol ester and *E coli* endotoxin, and CD18-independent for *S pneumoniae*; and 2) the mechanism of emigration in the lung may differ from that in the systemic circulation in response to the same stimulus, that is, CD18-dependent in the peritoneum versus CD18-independent in the lung with *S pneumoniae*.

Observations in a patient with severe CD11/CD18-deficiency suggest that CD18-independent pathways of neutrophil emigration in the lung may occur in man as well. Postmortem histologic examination of the lung demonstrated an accumulation of neutrophils in the alveoli, whereas neutrophils were not observed in extravascular tissue at sites of infection in the skin or intestine (Hawkins, H. and Anderson, D., personal communication).

One explanation for the CD18-independent mechanism of neutrophil emigration in the lung versus CD18-dependent in the peritoneum is that the pulmonary endothelium expresses different adhesion molecules than does the systemic endothelium. Although this is a reasonable possibility, at present there are no *in vitro* data to support this. An alternative explanation relates to the fact that the normal lung contains a significant population of alveolar macrophages. Mediators released by these cells in response to *S pneumoniae*

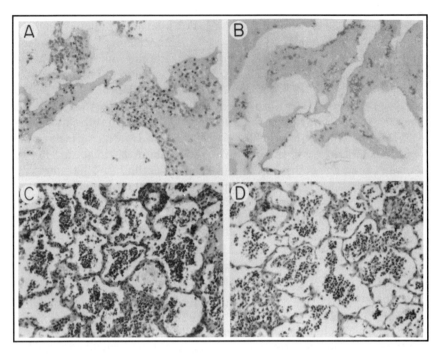

Figure 6-2. CD18-dependent and -independent mechanisms of neutrophil emigration. Neutrophil migration in response to *S pneumoniae*-induced inflammation was assessed in a control animal or an animal pretreated with the CD18 MAb 60.3 (2 mg/kg). *A*, *S pneumoniae*-containing sponge in control animal showing infiltration of sponge by neutrophils. *B*, *S pneumoniae*-containing sponge in animal pretreated with MAb 60.3, showing complete absence of neutrophils. *C*, lung of control animal after instillation of *S pneumoniae*, showing accumulation of neutrophils in alveoli. *D*, lung of MAb 60.3-treated animal (same animal as in *B*) following instillation of *S pneumoniae,* showing accumulation of neutrophils in alveoli. Adapted from Doerschuk, C.M., Winn, R.K., Coxson, H.O., and Harlan, J.M. (1990) *J. Immunol.* 144, 2327–2333 with permission.

could potentially induce expression of new adhesive ligands on the endothelium or activate different adhesion receptors on the leukocyte. This possibility is supported by the studies of Mileski et al.[64] who showed that the CD18-independent "phenotype" could be induced in normal peritoneum by the recruitment of macrophages. In these studies, neutrophil emigration into rabbit peritoneum was elicited by instillation of *E coli* or *S pneumoniae* organisms with or without pretreatment with the CD18 MAb 60.3. Peritoneal lavage was performed four hours after instillation of organisms and the number of neutrophils was determined. In the normal peritoneum, the CD18 MAb produced nearly 90% inhibition of neutrophil emigration. However, if the peritoneum was first "primed" by instillation of protease peptone 72 hours prior in order to elicit a macrophage-rich exudate, then the CD18 MAb only minimally inhibited *S pneumoniae*-induced emigration (36%), although it still inhibited *E coli* emigration by nearly 90%. If the "primed" peritoneum was washed to remove macrophages prior to instillation of *S pneumoniae* organisms, neutrophil emigration was again inhibited by nearly 90% by the CD18 MAb. Finally, instillation of macrophages obtained from protease peptone-treated animals into normal animals significantly reduced the inhibition produced by the CD18 MAb (48%). Overall, these results demonstrate that the CD18-independent mechanism of emigration that is observed in the pulmonary microcirculation in response to *S pneumoniae* organisms can be induced in the systemic microcirculation by maneuvers that augment the number of macrophages in the peritoneal cavity. The macrophage-generated product(s) elicited by *S pneumoniae* organisms and the adhesion molecules involved in this CD18-independent pathway remain to be identified.

ICAM-1

Intercellular adhesion molecule-1 (ICAM-1, CD54)[77,78] and ICAM-2[79] are ligands for CD11a/CD18. ICAM-1 is expressed at low levels on endothelium *in vivo*, and is up-regulated in response to inflammatory stimuli. ICAM-2 is constitutively expressed on endothelium. CD11a/CD18 recognizes both ICAM-1 and ICAM-2. Studies by Smith et al.[80] and by Diamond et al.[81] indicate that ICAM-1 is also a ligand for CD11b/CD18. Monoclonal antibodies to ICAM-1 have been demonstrated to inhibit lymphocyte and neutrophil emigration to tissues in several models of inflammation and immune reaction[68,70,82,83] (Table 6-1).

L-Selectin

The L(leukocyte)-selectin (LECAM-1, LAM-1) was first described in the mouse as the MEL-14 antigen, the "homing" receptor for lymphocyte binding to high endothelial venules of peripheral lymph nodes.[1] Lewinsohn et al.[84] showed that the MEL-14 antigen was also present on granulocytes and lymphocytes and that the MEL-14 MAb inhibited the binding of neutrophils and monocytes to inflamed high endothelial venules in tissue sections and at sites of acute inflammation in the skin. Subsequently, Jutila et al.[85] showed that MEL-14 also inhibited neutrophil accumulation in inflamed peritoneum. These observations using the MEL-14 MAb were confirmed by Watson et al.[86] using a soluble immunoglobulin chimera containing the murine homing receptor extracellular domain (LEC-IgG). Administration of LEC-IgG significantly decreased the number of neutrophils that migrated to the peritoneum in response to the inflammatory irritant thioglycolate. Watson

Table 6-1. Inhibitory Effect of MAbs to Adhesion Proteins on Leukocyte Emigration

Animal Model	MAb	Reference
Prevent neutrophil emigration into skin after intradermal injection of chemoattractants (rabbit)	CD18 (anti-β2)	Arfors et al.,[12] Lindbom et al.[14] Rampart & Williams[61] Nourshargh et al.[63]
Prevent neutrophil accumulation in intradermal polyvinyl sponge (rabbit)	CD18	Price et al.[45]
Prevent neutrophil emigration into CNS during bacterial meningitis (rabbit)	CD18	Tuomanen et al.[62]
Reduce neutrophil accumulation in skin in passive cutaneous anaphylatic reaction and reversed passive arthus reaction (rabbit)	CD18	Nourshargh et al.[63]
Reduce neutrophil emigration in inflamed peritoneum (rabbit)	CD18	Mileski et al.[64]
Prevent neutrophil emigration into myocardium following ischemia–reperfusion (primate)	CD18	Winquist et al.[65]
Reduce neutrophil emigration into inflamed lung (rabbit)	CD18	Doerschuk et al.[66] Horgan et al.[67] Barton et al.[68]
Reduce neutrophil accumulation in inflamed synovium (rabbit)	CD18	Jasin et al.[69]
Reduce neutrophil emigration following ischemia–reperfusion of intestine (cat)	CD18	Granger et al.[70]
Reduce neutrophil and monocyte accumulation in inflamed peritoneum (mouse)	CD11b (anti-Mac-1α)	Rosen et al.[74]
Prevent neutrophil accumulation in inflamed lung (mouse)	CD11b	Rosen[75]
Inhibit neutrophil and monocyte emigration to DTH in footpad (mouse)	CD11b	Rosen et al.[76]
Reduce neutrophil emigration into inflamed lung (rabbit)	CD54 (anti-ICAM-1)	Barton et al.[68]
Reduce T cell emigration into kidney during allograft rejection (primate)	CD54	Cosimi et al.[82]

Table 6-1. *(continued)*

Animal Model	MAb	Reference
Reduce eosinophil infiltration into lung following antigen challenge (primate)	CD54	Wegner et al.[83]
Inhibit neutrophil emigration following ischemia–reperfusion of intestine (cat)	CD54	Granger et al.[70]
Prevent neutrophil emigration into inflamed skin (mouse)	MEL-14 (anti-L-selectin)	Lewinsohn et al.[84]
Prevent neutrophil and monocyte emigration into inflamed ascites (mouse)	MEL-14 (anti-L-selectin)	Jutila et al.[85] Watson et al.[86]
Inhibit neutrophil emigration into inflamed peritoneum and lung (rat)	anti-ELAM-1 (anti-E-selectin)	Mulligan et al.[101]
Inhibit neutrophil influx into lung following antigen challenge (primate)	anti-ELAM-1 (anti-E-selectin)	Gundel et al.[102]
Inhibit lymphocyte emigration into cutaneous and joint inflammation (rat)	CD49d (anti-VLA-4α)	Issekutz[118]

and colleagues suggested that the homing receptor ligand that is expressed on peritoneal endothelium with inflammation is similar or identical to that in the lymph node high endothelium, and is presumably carbohydrate in nature.[86] The inhibitory effect of LEC-IgG or anti-LECAM-1 antibodies on neutrophil recruitment to sites of inflammation likely results from the ability of these reagents to inhibit leukocyte rolling, that is, rolling is a prerequisite for subsequent firm adhesion and emigration.[27,28]

E-Selectin

The E(endothelial)-Selectin, endothelial leukocyte adhesion molecule-1 (ELAM-1), is a cytokine-induced, endothelial-restricted selectin.[87] ELAM-1 mediates the binding of neutrophils,[88,89] monocytes,[90] and a subpopulation of lymphocytes[91–93] to cultured human endothelium stimulated by interleukin-1, lipopolysaccharide, or tumor necrosis factor. The carbohydrate counter-structure recognized by ELAM-1 appears to be the sialylated Lewis X antigen or a closely related structure.[94–97] ELAM-1 has been identified on endothelium at sites of inflammation in man, such as delayed type skin reactions,[98] rheumatoid synovium,[99] and allergic skin reactions.[100] Mulligan et al.[101] reported that an anti-ELAM-1 MAb reduced neutrophil accumulation in inflamed peritoneum and lung in the rat. Studies by Gundel et al.[102] showed that MAbs to ELAM-1 inhibited neutrophil accumulation in the lung in a primate model of antigen-induced inflammation.[98] Additional studies with blocking anti-ELAM-1 MAbs in animal models should define the role of this adhesion pathway in leukocyte emigration.

P-Selectin

The P(platelet)-selectin, granule membrane protein-140 (GMP-140)[103] or platelet activation-dependent granule-external membrane (PADGEM) protein,[104] now designated CD62, is a selectin receptor first detected on activated platelets and later on activated endothelial cells. In endothelial cells, CD62 is stored within Weibel-Palade bodies and is rapidly mobilized to the external plasma membrane upon stimulation by thrombin, histamine, or phorbol esters.[105] Once exposed on the lumenal surface, it is available for binding to leukocytes through interaction with a carbohydrate counter-structure on the leukocyte, the sialylated form of Lewis X.[106,107] CD62 has been localized to small veins and venules *in vivo*,[108] but an actual increase in luminal surface expression at sites of inflammation has not yet been demonstrated. Studies *in vivo* with blocking MAbs to CD62 have not yet been reported. However, as a rapidly mobilized endothelial selectin, CD62 may well be involved in leukocyte rolling. If so, blocking MAbs to CD62 would be expected to inhibit emigration.

VLA-4 and VCAM-1

Although neutrophil and monocyte emigration is markedly impaired in the CD11/CD18-deficiency syndrome, traffic of other leukocytes to extravascular tissues is only minimally affected. Biopsies of sites of infection in CD11/CD18-deficient patients exhibit an infiltration of lymphocytes, eosinophils, and occasional mononuclear phagocytes.[55] ELAM-1 and LECAM-1 may participate in the CD11/CD18-independent mechanism of lymphocyte emigration. However, the β_1 integrin receptor VLA-4 (CD49d/ CD29) and its recently identified endothelial ligand, vascular cell adhesion molecule-1 (VCAM-1),[109–111] also known as INCAM-110,[112] represent the best candidates for an alternative adhesion pathway for lymphocytes,[109–111,113,114] monocytes,[90,113] and eosinophils.[115–117] Preliminary studies by Issekutz[118] demonstrate that a CD49d (VLA-4α) MAb inhibited emigration of lymphocytes, primarily T memory cells, to sites of inflammation in the skin and joint.

GENETIC DEFICIENCY OF ADHESION PROTEINS

Studies of patients deficient in CD11/CD18 have provided important insights into the cell and molecular biology of leukocyte–endothelial adhesion.[53] Although the CD11/CD18-deficiency syndrome can be mimicked in experimental animals by infusion of a blocking CD18 MAb, this approach is limited in any model by the potential for binding of a MAb to trigger cellular responses in addition to blocking binding; that is, to transduce a signal, and by the inevitable development of endogenous antibodies to the foreign monoclonal antibody in chronic models. Animal models of CD11/CD18-deficiency are therefore of some interest. CD11/CD18-deficiency has been documented in a dog[119] and in cattle[120] with "granulocytopathy" syndromes. The clinical manifestations of CD11/CD18-deficiency in the affected animals are very similar to descriptions of CD11/CD18-deficient patients: neutrophilia, recurrent infections, impaired pus formation, delayed wound healing, and increased mortality.[3,53,58] Deficient expression of CD11/CD18 on the animals' leukocytes was confirmed by immunofluorescence studies. Unfortunately, the CD11/CD18-deficient dog has died, and no other affected dogs have been identified. Interestingly, the CD11/CD18-deficient cattle belong to a kindred involving the same sire three times in the pedigree. Occurrence of CD11/CD18-deficiency may therefore occur more frequently in cattle than in dogs because of inbreeding and use of artificial insemination. It will be important to determine whether "granulocytopathy" syndromes produced by CD11/CD18-deficiency occur in smaller laboratory animals that are more suitable for *in vivo* studies. Alternatively, CD11/CD18-deficiency could potentially be produced in laboratory animals using the technique of homologous recombination to "knock-out" the β_2 gene. This genetic approach could be applied to other adhesion proteins as well.

Anti-Adhesion Therapy

Leukocyte traffic across the vessel wall to extravascular vascular tissue is necessary for host defense against microbial organisms or foreign antigens and repair of tissue damage. Under some circumstances, however, leukocyte–endothelial interactions may have deleterious consequences for the host. During the process of adherence and transendothelial migration, leukocytes may release products such as oxidants, proteases, or cytokines that directly damage endothelium or cause endothelial dysfunction.[121] Subsequently, emigrated leukocytes may contribute to tissue damage by releasing a variety of inflammatory mediators. Finally, sticking of single leukocytes within the capillary lumen or aggregation of leukocytes within larger vessels may lead to microvascular occlusion and may produce ischemia. Leukocyte-mediated vascular and tissue injury has been implicated in the pathogenesis of a wide variety of clinical disorders, including acute and chronic allograft rejection; vasculitis; rheumatoid arthritis; inflammatory skin diseases; adult respiratory distress syndrome; and ischemia-reperfusion syndromes (e.g., myocardial infarction, shock, stroke, organ transplantation, crush injury, and limb replantation).

Inhibition of leukocyte adherence to endothelium—"anti-adhesion" therapy—represents a novel approach to the treatment of those inflammatory and immune disorders in which leukocytes contribute significantly to vascular and tissue injury. Studies *in vitro* indicate that close approximation of the leukocyte to the endothelial cell is required for leukocyte-mediated damage to occur. Firm adhesion of the leukocyte to the endothelial cell forms a protected microenvironment at the interface of the leukocyte and endothelial cell that is inaccessible to plasma inhibitors.[121] Highly reactive oxidants, proteases, and phospholipase products released by adherent leukocytes at the interface can react with and damage the endothelium. Inhibition of firm leukocyte adherence by a MAb to CD18, for example, prevents formation of a protected microenvironment, and thereby reduces this type of "innocent bystander" injury to endothelium (Figure 6-3). Inhibition of leukocyte adherence to endothelium will also prevent emigration to tissue, and, consequently, reduce tissue damage produced by emigrated leukocytes. Finally, inhibition of leukocyte adherence to endothelium or homotypic aggregation will prevent microvascular occlusion.

Efficacy

The potential for MAbs to adhesion proteins to reduce vascular and tissue damage has been examined in a variety of experimental models (Table 6-2). Striking protective effects have been observed in an amazing spectrum of inflammatory and immune reactions, including: skin lesions,[12,14,63,76] bacterial meningitis,[62] endotoxic shock,[122] burns,[123] aspiration pneumonitis,[124] arthritis,[69] nonsteroidal induced gastric injury,[125] autoimmune diabetes,[126] nerve degeneration,[127] allograft rejection,[82,128] allergic asthma,[83,102] oxygen toxicity,[129] and acute lung inflammation.[101] Notably, a CD54 (anti-ICAM-1) MAb is now in a phase 1 clinical trial for the prevention of renal allograft rejection (R. Faanes, Boeringer-Ingelheim, Ridgefield, CT, personal communication, June 1991).

Anti-adhesion therapy has also been tested successfully in models of ischemia–reperfusion injury. While it is clear that prolonged ischemia will produce damage to tissue as a result of anoxia, paradoxically, further injury occurs following reperfusion. At least in some settings, this reperfusion injury appears to involve the inflammatory response.[130] The precise signals (e.g., oxidants, activated complement components, phospholipase products, or cytokines) that initiate the inflammatory response during ischemia–reperfusion

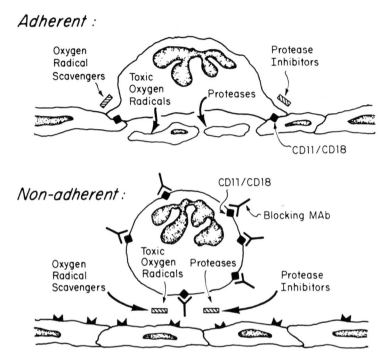

Figure 6-3. Role of neutrophil adherence in neutrophil-mediated vascular injury. Tight neutrophil adherence to endothelial cells via CD11/CD18 forms a protected microenvironment that is not accessible to plasma inhibitors. Inhibition of neutrophil adherence by a CD11b or CD18 MAb prevents formation of the protected microenvironment and thereby reduces vascular injury. Adapted from Harlan, J.M., (1987) Acta Med. Scand. Suppl. 715, 123–129, with permission.

have not been identified and may vary with the particular organ or disorder. However, early studies utilizing neutrophil depletion indicated that neutrophils may contribute significantly to reperfusion damage by generating oxidants or releasing proteases that damage the microvasculature or adjacent tissue, or by plugging capillaries producing further ischemia[131–135] (Figure 6-4).

Hernandez et al.[136] were the first to demonstrate the critical role of leukocyte adhesion in ischemia–reperfusion injury. They showed that pretreatment with a CD18 MAb prevented an increase in microvascular permeability following ischemia–reperfusion of the cat intestine. Simpson and colleagues[137,138] subsequently showed that a CD11b MAb reduced infarct size in a canine model of myocardial ischemia–reperfusion, a finding recently confirmed in primate[65] and feline[139] models.

Vedder et al.[140] examined the role of leukocyte adhesion in a rabbit ear model of isolated tissue ischemia–reperfusion. They reported that a CD18

Table 6-2. Anti-Adhesion Therapy in Experimental Models of Vascular and Tissue Injury

Animal Model	MAb	Reference
A. Inflammatory/Immune		
Reduce permeability edema in inflammatory skin lesions (rabbit)	CD18 (anti-β_2)	Arfors et al.[12] Lindbom et al.[14]
Reduce brain edema and death produced by bacterial meningitis (rabbit)	CD18	Tuomanen et al.[62]
Reduce tissue edema associated with delayed-type hypersensitivity reactions (rabbit, mouse)	CD18	Lindbom et al.[14] Rosen et al.[76]
Protect against vascular injury and death in endotoxic shock (rabbit)	CD18	Thomas et al.[122]
Prevent second degree burns from becoming third degree burns (rabbit)	CD18	Bucky et al.[123]
Reduce remote lung injury following aspiration (rabbit)	CD18	Goldman et al.[124]
Amelioration of antigen-induced arthritis (rabbit)	CD18	Jasin et al.[69]
Prevent indomethacin-induced gastric injury (rabbit)	CD18	Wallace et al.[125]
Inhibit development of auto-immune diabetes (mouse)	CD11b (anti-Mac-1α)	Hutchings et al.[126]
Prevent Wallerian degeneration following section of sciatic nerve (mouse)	CD11b	Lunn et al.[127]
Attenuate renal allograft rejection (primate)	CD54 (anti-ICAM-1)	Cosimi et al.[82]
Reduce airway hyper-responsiveness in allergic asthma (primate)	CD54	Wegner et al.[83]
Attenuate lung damage and dysfunction secondary to oxygen toxicity (mouse)	CD54	Wegner et al.[129]
Reduce remote lung injury following aspiration (rabbit)	CD54	Goldman et al.[124]
Prolong cardiac allograft survival (primate)	CD54	Flavin et al.[128]

Table 6-2. *(continued)*

Animal Model	MAb	Reference
Reduce late-phase bronchoconstriction following antigen challenge (primate)	anti-ELAM-1 (anti-E-selectin)	Gundel et al.[102]
Reduce permeability edema in acute lung inflammation (rat)	anti-ELAM-1 (anti-E-selectin)	Mulligan et al.[101]
B. Ischemia–Reperfusion (I/R)		
Reduce permeability edema following I/R of intestine (cat)	CD18	Hernandez et al.[136]
Reduce edema and tissue drainage following I/R of ear (rabbit)	CD18	Vedder et al.[140]
Reduce vascular and tissue damage following hemorrhagic shock and resuscitation (rabbit, primate)	CD18	Vedder et al.[143,144] Mileski et al.[146]
Reduce myocardial damage following I/R (primate, cat)	CD18	Winquist et al.[65] Ma et al.[139]
Reduce lung injury following I/R of hindlimb (rabbit)	CD18	Welbourn et al.[147]
Reduce lung injury following I/R (rabbit)	CD18	Horgan et al.[67] Bishop et al.[148]
Reduce permeability edema in skeletal muscle following I/R (rabbit)	CD18	Carden et al.[149]
Reduce central nervous system damage following I/R of spinal cord (rabbit)	CD18	Clark et al.[150]
Reduce edema and tissue damage following frostbite and rewarming (rabbit)	CD18	Mileski et al.[151]
Reduce infarct size following I/R of myocardium (dog)	CD11b	Simpson et al.[137,138]
Reduce permeability edema following I/R of intestine (cat)	CD54	Granger et al.[70]

MAb markedly attenuated neutrophil accumulation and associated vascular and tissue injury produced by 10 hours of warm ischemia followed by reperfusion. Importantly, they found that protection was equivalent when the MAb was administered just prior to reperfusion, but following ischemia,

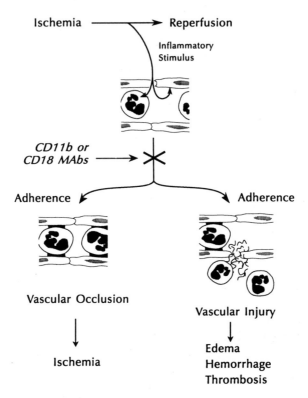

Figure 6-4. Potential mechanisms of action of CD11b or CD18 MAbs in reducing vascular and tissue injury following ischemia–reperfusion.

as it was when administered prior to ischemia (Figure 6-5). These results indicated that the leukocyte-mediated injury was in fact a reperfusion injury and that it could be significantly attenuated by a CD18 MAb administered at the time of reperfusion.

This approach was further tested by Vedder and colleagues in a rabbit model of hemorrhagic shock and resuscitation that may be relevant to clinical multiple organ failure syndrome.[141,142] They hypothesized that shock (ischemia) and subsequent resuscitation (reperfusion) triggered neutrophil adhesion throughout the microcirculation and led to the development of multiple organ failure. Administration of a CD18 MAb prior to[143] shock or at the time of resuscitation[144] resulted in a dramatic reduction in vascular and tissue injury and a marked improvement in survival (Figure 6-6). Consistent with these results obtained with the CD18 MAb, Barroso-Aranda et al.[145] found that prior neutrophil depletion prevented death in a rat model of hemorrhagic shock. They observed more frequent no-reflow in the capillaries of control animals due to occlusion by trapped neutrophils.

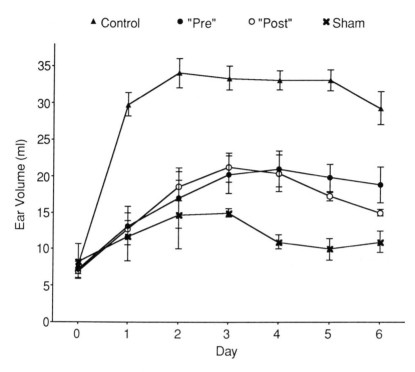

Figure 6-5. Ischemia–reperfusion injury of rabbit ear. The CD18 MAb 60.3 attenuates edema formation after 10 hours of ischemia and then reperfusion. Days after reperfusion are indicated on the abscissa and ear volume is indicated on the ordinate. Edema was quantified by volume displacement. Mean values for sham-operated animals (n=3) are represented by an "x." Values of control-ischemic animals (n=8) are represented by closed triangles. Values for animals that received CD18 MAb prior to ischemia and again prior to reperfusion ("Pre;" n=5) are represented by closed circles. Values for animals that received CD18 MAb only prior to reperfusion ("Post;" n=5) are represented by open circles. Error bars represent ±SEM. Ear volumes in both the CD18 MAb-pretreated and CD18 MAb-posttreated animals were significantly less than in control-ischemic animals ($P<0.005$ by analysis of variance), but not significantly different from sham-operated animals ($P>0.05$). Ear volumes in the CD18 MAb-pretreated and CD18 MAb-posttreated animals were not different from each other ($P>0.05$). Reproduced from Vedder, N.B., Winn, R.K., Rice, C., Chi, E.Y., Arfors, K.E., and Harlan, J.M. (1990) *Proc. Natl. Acad. Sci. USA* 87, 2643–2646 with permission.

The importance of neutrophil adhesion in the pathophysiology of hemorrhagic shock and resuscitation was recently confirmed by Mileski and colleagues[146] in a primate model of hemorrhagic shock and resuscitation. Control animals developed generalized microvascular injury manifested by diffuse capillary leak and required massive volumes of fluid to sustain cardiac output (Figure 6-7). In contrast, animals treated with a single injection of a

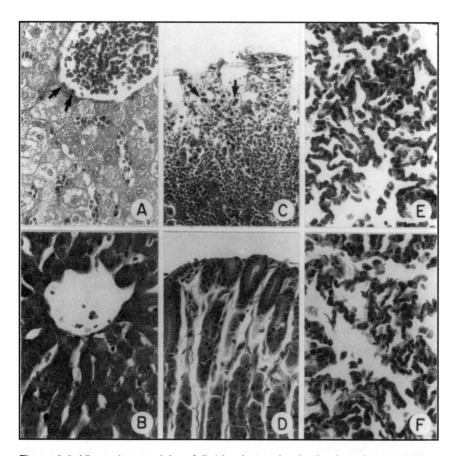

Figure 6-6. Visceral organ injury following hemorrhagic shock and resuscitation. Histologic comparison of organ injury in rabbits subjected to two hours of hemorrhagic shock, then treated with saline solution or the CD18 MAb 60.3 immediately before resuscitation. For this comparison, animals were killed 24 hours following shock, specimens fixed immediately, stained with hematoxylin and eosin, and photographed at × 132 magnification. A, section through liver of control rabbit shows hepatocyte necrosis with neutrophils adherent to the endothelium and infiltrating the tissue (*arrows*). B, section through liver of identically injured rabbit, treated with MAb 60.3, shows no organ injury. C, severe gastric mucosal injury of a control rabbit. The lumenal epithelium is completely destroyed, and there is evidence of diffuse hemorrhage and neutrophil infiltration (*arrows*). D, normal epithelial stomach lining of MAb 60.3-treated rabbit. E, section through lung of control rabbit showing atelectasis, leukocyte infiltration, and hemorrhage. F, lung of MAb 60.3-treated rabbit. Cellular infiltration exists as in control rabbit, with only slightly less hemorrhage. Reproduced from Vedder, N.B., Fouty, B.W., Winn, R.K., Harlan, J.M., and Rice, C.L. (1989) *Surgery* 10, 509–516 with permission.

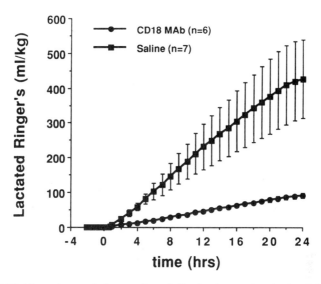

Figure 6-7. Fluid requirements for monkeys following hemorrhagic shock. Treatment with saline in the CD18 MAb 60.3 was given at the time of resuscitation in both groups of animals and consisted of 4 ml/kg plus volume sufficient to sustain baseline cardiac output. Adapted from Mileski, W.J., Winn, R.K., Vedder, N.B., Pohlman, T.H., Harlan, J.M., and Rice, C.L. (1990) *Surgery* 108, 206–212 with permission.

CD18 MAb at the time of resuscitation required no fluids above maintenance to maintain cardiac output. In addition, all of the control animals developed visceral injury (gastritis) and two of five control animals died, whereas none of the CD18 MAb-treated animals developed gastritis or died.

Anti-adhesion therapy has also been tested in animal models of ischemia–reperfusion injury of lung,[67,147,148] skeletal muscle,[149] central nervous system,[150] and frostbite.[151] In the latter model, it was hypothesized that freezing of tissue produces ischemia and rewarming induces reperfusion. Importantly, only a 30-minute delay in administration of the CD18 MAb following rewarming significantly reduced efficacy (Figure 6-8). This result suggests that, at least in this model, the critical interaction of leukocytes with endothelium occurs very early after reperfusion.

Safety

These studies of anti-adhesion therapy in the various experimental models demonstrate that inhibition of leukocyte adhesion to endothelium can significantly reduce vascular and tissue damage. However, as clearly demonstrated by the CD11/CD18-deficient patients, neutrophil adherence to endothelium is required for the eradication of bacterial infections at extravascular sites. Thus, there is genuine cause for concern that even transient inhibition of

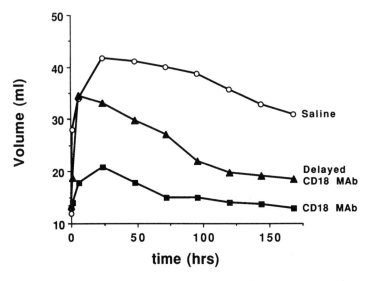

Figure 6-8. Rabbit limb volume measured by displacement following severe cold exposure (–15°C for 30 minutes) in three groups of animals. The saline (*open circle*) and CD18 MAb group (*closed square*) received treatment at the beginning of rewarming and the delayed CD18 MAb 60.3 group (*closed triangle*) received treatment 30 minutes after rewarming.

neutrophil adherence to endothelium will result in a marked increase in susceptibility to and severity of bacterial infection. Indeed, Rosen and coworkers[152] reported that pretreatment of mice with a CD11b MAb markedly increased mortality following inoculation with *Listeria monocytogenes*, illustrating the potential risks of anti-adhesion therapy as a therapeutic strategy. In contrast, however, Tuomanen et al.[62] found that administration of a CD18 MAb reduced mortality in a rabbit model of bacterial meningitis by preventing disruption of the blood–brain barrier and resultant cerebral edema. These authors noted that the course of bacterial growth in the blood and the cerebral spinal fluid was not adversely affected by the CD18 MAb. Similarly, Mileski et al.[153] reported that a CD18 MAb did not increase mortality or infectious complications in a rabbit model of bacterial peritonitis (Figure 6-9). More recently, Sharar and colleagues[154] showed that a CD18 MAb did not increase the number or extent of abscesses resulting from clinically relevant inocula of staphylococcal organisms in the skin of rabbits (Figure 6-10). At much higher inocula, however, extensive, severe infections were observed in the CD18 MAb-treated animals, even with appropriate antibiotic coverage.

Overall, these studies suggest that it may be safe to completely inhibit neutrophil adhesion for a limited period of time, even in patients at high risk of infection or with certain types of established infection. Obviously, though, anti-adhesion therapy is a two-edged sword that must be handled with great caution.

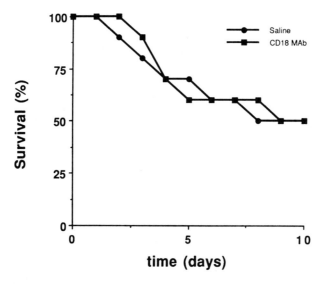

Figure 6-9. Survival in rabbits following devascularization of the appendix to induce sepsis. An appendectomy was performed on day zero and there were 10 animals in each group. Rabbits received treatment with saline or the CD18 MAb 60.3 at the time of devascularization and again at appendectomy. Adapted from Mileski, W.J., Winn, R.K., Harlan, J.M., and Rice, C.L. (1991) *Surgery* 109, 497–501, with permission.

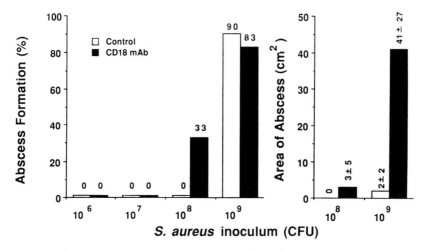

Figure 6-10. Percent of abscess formation and area of abscess following subcutaneous injection of *S aureus* bacteria. Rabbits were treated at the time of the injection with either saline or the CD18 MAb 60.3. There were no abscesses at the lower inocula (10^6 and 10^7 CFU) in either group. Adapted from Sharar, S.R., Winn, R.K., Murry, C.E., Harlan, J.M., and Rice, C.L. (1991) *Surgery* 110, 213–220 with permission.

Future Directions

Although the administration of MAbs (murine or humanized), soluble receptors, or peptide antagonists is feasible in acute settings such as myocardial infarction or shock, or even in subacute settings such as allograft rejection, this approach may not be practical for extended treatment in chronic disorders such as rheumatoid arthritis. By identifying the adhesion pathways involved in the recruitment of critical leukocyte subpopulations to target organs in chronic disorders, it may be possible to develop small molecules (i.e., carbohydrates) that block receptor–ligand interactions or that inhibit the induction or activation of the relevant adhesion molecules.

CONCLUSION

The adhesive interaction between leukocytes and endothelium is a critical event in host defense and repair. Leukocyte adherence to endothelium also contributes to the pathologic consequences of inflammatory and immune reactions. *In vitro* studies have identified and characterized a number of putative adhesion proteins, suggested potential cellular interactions, and generated reagents for testing *in vivo*. The contribution of several of the adhesion proteins identified *in vitro* to leukocyte adherence and emigration has been confirmed in animal models of inflammation and immune reaction. Studies in animal models of disease have also demonstrated that inhibition of leukocyte adherence to endothelium—"anti-adhesion therapy"—can markedly reduce vascular and tissue injury. This synergy between *in vitro* and *in vivo* studies has led to rapid, dramatic progress in our understanding of the cell biology, physiology, and clinical relevance of leukocyte–endothelial adhesion, and augurs an exciting, new approach to the therapy of a diverse array of important clinical disorders.

REFERENCES

1. Gallatin, W.M., Weissman, I.L., and Butcher, E.C. (1983) *Nature* 304, 30–34.
2. Springer, T.A. (1990) *Nature* 346, 425–434.
3. Anderson, D.C. and Springer, T.A. (1987) *Ann. Rev. Med.* 38, 175–194.
4. Berg, E.L., Goldstein, L.A., Jutila, M.A., Nakache, M., Picker, L.J., Streeter, P.R., Wu, N.W., Zhou, D., and Butcher, E.C. (1989) *Immunological Rev* 108, 5–18.

Acknowledgments: We thank our many colleagues who contributed to the authors' studies described in this chapter, particularly Patrick Beatty, William Mileski, Timothy Pohlman, Samuel Sharar, Karl E. Arfors, and James Hogg.

5. Grant, L. (1973) in *The Inflammatory Process* (Zweifach, B, Grant, L., and McCluskey, L., eds.) pp. 205–249, Academic Press, Orlando, Florida.
6. Dutrochet, M.H. (1824) *Recherches, Anatomiques et Physiologiques sur la Structure Intime des Animaux et des Vegetaux, et sur Leur Motilite.* Bailliere et Fils, Paris.
7. Movat, H.Z. (1989) *Pathol Immunopathol Res.* 8, 35–41.
8. Cohnheim, J. (1889) *Lectures in General Pathology,* Vol. 1., pp. 242–382, New Sydenham Society, London.
9. Metchnikoff, E. (1893, republished 1968). *Lectures on the Comparative Pathology of Inflammation.* Paul, Kegan, London; republ. pp. 1–218, Dover, New York.
10. Clark, E.R. and Clark, E.L. (1935) *Am. J. Anat.* 57, 385–438.
11. Allison, F., Jr., Smith, M.R., and Wood, W.B., Jr. (1955) *J. Exp. Med.* 102, 655–668.
12. Arfors, K.E., Lundberg, D., Lindbom, L., Lundberg, K., Beatty, P.G., and Harlan, J.M. (1987) *Blood* 69, 338–340.
13. Beatty, P.G., Ledbetter, J.A., Martin, P.J., Price, T.H., and Hansen, J.A. (1983) *J. Immunol.* 31, 2913–2918.
14. Lindbom, L., Lundberg, C., Prieto, J., Raud, J., Nortamo, P., Gahmberg, C.G., and Patarroyo, M. (1990) *Clin. Immunol. Immunopath.* 57, 105–119.
15. Argenbright, L.W., Letts, L.G., and Rothlein, R. (1991) *J. Leukocyte Biol.* 49, 253–257.
16. Perry, M.A. and Granger D.N. (1991) *J. Clin. Invest.* 87, 1798–1804.
17. Lawrence, M.B., McIntire L.V., and Eskin, S.G. (1987) *Blood* 70, 1284–1290.
18. Lawrence, M.B., Smith, C.W., Eskin, S.G., and McIntire, L.V. (1990) *Blood* 75, 227–237.
19. Worthen, G.S., Smedly, L.A., Tonnesen, M.G., Ellis, D., Voelkel, N.F., Reeves, J.J., and Henson, P.M. (1987) *J. Appl. Physiol.* 63, 2031–2041.
20. Kishimoto, T.K., Jutila, M.A., Berg, E.L, and Butcher, E.C. (1989) *Science* 245, 1238–1241.
21. Fiebig, E., Ley, K. and Arfors, K. E. (1991) *Int. J. Microcirc. Clin. Exp.* 10, 127–144.
22. Atherton, A. and Born G.V.R. (1972) *J. Physiol.* 222, 447–474.
23. Atherton, A. and Born, G.V.R. (1973) *J. Physiol.* 233, 157–165.
24. Mayrovitz, H.N., Tuma, R.F, and Weideman, M.P. (1980) *Microvasc. Res.* 20, 264–274.
25. Atherton, A. and Born, G.V.R. (1973) *J. Physiol.* 234, 66–67.
26. Ley, K., Cerrito, M., and Arfors, K.E. (1991) *Am. J. Physiol.* 260, H1673–1991.
27. Ley, K., Gaehtgens, P., Fennie, D., Singer, M.S., Lasky, L.A., and Rosen, S.D. (1991) *Blood* 77, 2553–2555.
28. von Andrian U.H., Chambers, J.D., McEvoy, L.M., Bargatze, R.F., Arfors, K.E., and Butcher, E.C. (1991) *Proc. Natl. Acad. Sci. USA* 88, 7538–7542
29. Doerschuk, C.M., Allard, M.F., Martin, B.A., MacKenzie, A., Autor, A.P., and Hogg, J.C. (1987) *J. Appl. Physiol.* 63, 1806–1815.
30. Van Furth, R. and Sluiter, W. (1986) *J. Exp. Med.* 163, 474–479.
31. Ohgami, M., Doerschuk, C.M., Gie, R.P., English, D., and Hogg, J.C. (1991) *J. Appl. Physiol.* 70, 152–157.

32. Doerschuk, C.M., Downey, G.P., Doherty, D.E., English, D., Gie, R.P., Ohgami, M., Worthen, G.S., Henson, P.M., and Hogg, J.C. (1990) *J. Appl. Physiol.* 68, 1956–1961.
33. Pabst, R., Binns, R.M., Licence, S.T., and Peter, M. (1987) *Am. Rev. Respir. Dis.* 136, 1213–1218.
34. Lien, D.C., Wagner, W.W., Jr., Capen, R.L., Haslett, C., Hanson, W.L., Hofmeister, S.E., Henson, P.M., and Worthen, G.S. (1987) *J. Appl. Physiol.* 62, 1236–1243.
35. Lien, D.C., Worthen, G.S., Capen, R.L., Hanson, W.L., Checkley, L.L., Janke, S.J., Henson, P.M., and Wagner, W.W. (1990) *Am. Rev. Resp. Dis.* 141, 953–959.
36. Thommasen, H.V., Martin, B.A., Wiggs, B.R., Quiroga, M., Baile, E.M., and Hogg, J.C. (1984) *J. Appl. Physiol.* 56, 966–974.
37. Downey, G.P., Doherty, D.E., Schwabb, B., III, Elson, E.L., Henson, P.M., and Worthen, G.S. (1990) *J. Appl. Physiol.* 69, 1767–1778.
38. Schmid-Shonbein, G.W., Shih, Y.Y., and Chien, S. (1985) *Blood* 56, 866–875.
39. Guntheroth, W.G., Luchtel, D.L., and Kawabori, I. (1982) *J. Appl. Physiol.* 53, 510–515.
40. Hogg, J.C., McLean, T., Martin, B.A., and Wiggs, B. (1988) *J. Appl. Physiol.* 65, 1217–1225.
41. Chien, S. (1987) *Ann. Rev. Physiol.* 49, 177–192.
42. Bierman, H.R., Kelly, K.H., Cordes, F.L., Byron, R.L., Polhemus, J.A., and Rapproport, S. (1952) *Blood* 7, 683–692.
43. Buchanan, M.R., Crowley, C.A., Rosein, R.E., Gimbrone, M.A., Jr., and Babior, B.M. (1982) *Blood* 60, 160–165.
44. Nunoi, H., Yanabe, Y., Higuchi, S., Tsuchiya, H., Yamamoto, J., Matsuda, I., Naito, M., Takahashi, K., Fujita, K., Uchida, M., Kobayashi, K., Jono, M., and Malech, H. (1988) *Human Pathol.* 19, 753–759.
45. Price, T.H., Beatty, P.G., and Corpuz, S.R. (1987) *J. Immunol.* 139, 4174–4177.
46. Yoder, M.C., Checkley, L.L., Giger, U., Hanson, W.L., Kirk, K.R., Capen, R.L., and Wagner, W.W., Jr. (1990) *J. Appl. Physiol.* 69, 207–213.
47. Lundberg, C. and Wright, S.D. (1990) *Blood* 76, 1240–1245.
48. Doerschuk, C.M., English, D., Harlan, J.M., and Hogg, J.C. (1990) *FASEB J.* 4, A496.
49. Yong, K., Addison, I.E., Johnson, B., Webster, A.D.B., and Linch, D.C. (1991) *Br. J. Haematol.* 77, 150–157.
50. Doerschuk, C.M., Allard, M.F., and Hogg, J.C. (1989) *J. Appl. Physiol.* 67, 88–95.
51. Worthen, G.S., Schwab, B., III, Elson, E.L., and Downey, G.P. (1989) *Science* 245, 183–185.
52. Harlan, J.M. (1985) *Blood* 65, 513–525.
53. Schwartz, B.A., and Harlan, J.M. in *Monographs on Cell and Tissue Physiology*, No. 17 (Gordon, J.L., ed.) Elsevier Science Publishers BV, Amsterdam, The Netherlands. In press.
54. Crowley, C.A., Curnutte, J.T., Rosin, R.E., Andre-Schwartz, J., Gallin, J.I., Klempner, M., Snyderman, R., Southwick, F.S., Stossel, T.P., and Babior, B.M. (1980) *N. Engl. J. Med.* 302, 1163–1168.
55. Anderson, D.C., Schmalstieg, F.C., Finegold, M.J., Hughes, B.J., Rothlein, R., Miller, L.J., Kohl, S., Tosi, M.F., Jacobs, R.L., Waldrop, T.C., Goldman, A.S., Shearer, W.T., and Springer, T.A., (1985) *J. Infect. Dis.* 152, 668–689.

56. Bowen, T.J., Ochs, H.D., Altman, L.C., Price, T.H., Van Epps, D.E., Brautigan, D.L., Rosin, R.E., Perkins, W.D., Babior, B.M., Klebanoff, S.J., and Wedgwood, R.J. (1982) *J. Pediatr.* 101, 932–940.
57. Waldrop, T.C., Anderson, D.C., Hallmon, W.W., Schmalstieg, F.C., and Jacobs, R.L. (1986) *J. Periodontol.* 58, 400–416.
58. Todd, R.F., III, and Freyer, D.R. (1988) *Hematol. Oncol. Clin. North Am.* 2, 13–31.
59. Beatty, P.G., Ochs, H.D., Rosen, H., Hansen, J.A., Harlan, J.M., Price, T.H., Taylor, R.F., and Klebanoff, S.J. (1984) *Lancet* 1, 535–537.
60. Harlan, J.M., Killen, P.D., Senecal, F.M., Schwartz, B.R., Yee, E.K., Taylor, R.F., Beatty, P.G., Price, T.H., and Ochs, H.D. (1985) *Blood* 66, 167–178.
61. Rampart, M. and Williams J.T. (1988) *Br. J. Pharmacol.* 94, 1143–1148.
62. Tuomanen, E.I., Saukkonen, K., Sande, S., Cioffe, C., and Wright, S.D. *J. Exp. Med.* (1989) 170, 959–968.
63. Nourshargh, S., Rampart, M., Hellewell, P.G., Jose, P.J., Harlan, J.M., Edwards, A.J., and Williams, T.J. (1989) *J. Immunol.* 142, 3193–3198.
64. Mileski, W., Harlan, J., Rice, C., and Winn, R. (1990) *Circ. Shock* 31, 259–267.
65. Winquist, R., Frei, P., Harrison, P., McFarland, M., Letts, G., Van, G., Andrews L, Rothlein, R., and Hintze, T. (1990) *Circulation* 82, III–701.
66. Doerschuk, C.M., Winn, R.K., Coxson, H.O., and Harlan, J.M. (1990) *J. Immunol.* 144, 2327–2333.
67. Horgan, M.J., Wright, S.D., and Malik, A.B. (1990) *Am. J. Physiol.* 259, L315–L319.
68. Barton, R.W., Rothlein, R., Ksiazek, J., and Kennedy C. (1989) *J. Immunol.* 143, 1278–1282.
69. Jasin, H.E., Lightfoot, E., Kavanaugh, A., Rothlein, R., Faanes, R.B., and Lipsky, P.E. (1990) *Arthritis Rheum.* 33, S34.
70. Granger, D.N., Russell, J., Arfors, K.E., Rothlein, R., and Anderson, D.C. (1991) *FASEB J.* 5, A1753.
71. Forsyth, K.D. and Levinsky, R.J. (1989) *Clin. Exp. Immunol.* 75, 265–268.
72. Smith, C.W., Marlin, S.D., Rothlein, R., Toman, C., and Anderson, D.C. (1989) *J. Clin. Invest.* 83, 2008–2017.
73. Lo, S.K., van Seventer, G.A., Levin, S.M., and Wright, S.D. (1989) *J. Immunol.* 143, 3325–3329.
74. Rosen, H. and Gordon, S. (1987) *J. Exp. Med.* 1685–1701.
75. Rosen, H. (1990) *J. Leukocyte Biol.* 48, 465–469.
76. Rosen, H., Milon, G., and Gordon, S. (1989) *J. Exp. Med.* 169, 535–548.
77. Rothlein, R., Dustin, M.L., Marlin, S.D., and Springer, T.A. (1986) *J. Immunol.* 137, 1270–1274.
78. Staunton, D.E., Marlin, S.D., Stratowa, C., Dustin, M.L., and Springer, T.A. (1988) *Cell* 52, 925–933.
79. Staunton, D.E., Dustin, M.L., and Springer, T.A. (1989) *Nature* 339, 61–64.
80. Smith, C.W., Rothlein, R., Hughes, B.J., Mariscalco, M.M., Rudloff, H.E., Schmalsteig, F.C., and Anderson, D.C. (1988) *J. Clin. Invest.* 82, 1746–1756.
81. Diamond, M.S., Staunton, D.E., de Fougerolles, A.R., Stacker, S.A., Garcia-Aguilar, J., Hibbs, M.L., and Springer, T.A. (1990) *J. Cell Biol.* 111, 3129–3139.
82. Cosimi, A.B., Conti, D., Delmonico, F.L., Preffer, F.I., Wee, S.L., Rothlein, R., Faanes, R., and Colvin, R.B. (1990) *J. Immunol.* 144, 4604–4612.

83. Wegner, C.D., Gundel, R.H., Reilly, P., Haynes, N., Letts, L.G., and Rothlein, R. (1990) *Science* 247, 456–459.
84. Lewinsohn, D.M., Bargatze, R.F., and Butcher, E.C. (1987) *J. Immunol.* 138, 4313–4321.
85. Jutila, M.A., Rott, L., Berg, E.L., and Butcher, E.C. (1989) *J. Immunol.* 143, 3318–3324.
86. Watson, S.R., Fennie, C., and Lasky, L.A. (1991) *Nature* 349, 164–167.
87. Bevilacqua, M.P., Stengelin, S., Gimbrone, M.A., Jr., and Seed, B. (1989) *Science* 243, 1160–1165.
88. Bevilacqua, M.P., Pober, J.S., Mendrick, D.L., Cotran, R.S., and Gimbrone, M.A., Jr. (1987) *Proc. Natl. Acad. Sci. USA* 84, 9238–9242.
89. Bevilacqua, M.P., Pober, J.S., Wheeler, M.E., Cotran, R.S., and Gimbrone, M.A., Jr. (1985) *J. Clin. Invest.* 76, 2003–2011.
90. Carlos, T., Kovach, N., Schwartz, B., Osborn L., Rosa, M., Newman, B., Wayner, E., L., Lobb, R., and Harlan, J.M. (1991) *Blood* 77, 2266–2271.
91. Graber, N., Gopal, T.V., Wilson, D., Beall, L.D., Polte, T., and Newman, W. (1990) *J. Immunol.* 145, 819–830.
92. Picker, L.J., Kishimoto, T.K., Smith, C.W, Warnock, R.A., and Butcher, E.C. (1991) *Nature* 349, 796–799.
93. Shimizu, Y., Shaw, S., Graber, N., Gopal, T.V., Horgan, K.J., van Seventer, G.A., and Newman, W. (1991) *Nature* 349, 799–802.
94. Phillips, M.L., Nudelman, E., Gaeta, F.C.A., Perez, M., Singhal, A.K., Hakomori, S.I., and Paulson, J.C. (1990) *Science* 250, 1130–1131.
95. Walz, G., Aruffo, A., Kolanus, W., Bevilacqua, M., and Seed, B. (1990) *Science* 250, 1132–1135.
96. Tiemeyer, M., Swiedler, S.J., Ishihara, M., Moreland, M., Schweingruber, H., Hirtzer, P., and Brandley, B.K. (1991) *Proc. Natl. Acad. Sci. USA* 88, 1138–1142.
97. Lowe, J.B., Stoolman, L.M., Nair, R.P., Larsen, R.D., Berhend, T.L., and Marks, R.M. (1990) *Cell* 63, 475–484.
98. Cotran, R.S., Gimbrone, M.A., Jr., Bevilacqua, M.P., Mendrick, D.L., and Pober, J.S. (1986) *J. Exp. Med.* 164, 661–666.
99. Koch, A.E., Burrows, J.C., Haines, G.K., Carlos, T.M., Harlan, J.M., and Leibovich, S.J. (1991) *Lab. Invest.* 64, 313–320.
100. Leung, D.Y.M., Pober, J.S., and Cotran, R.S. (1991) *J. Clin. Invest.* 87, 1805–1809.
101. Mulligan, M.S., Varani, J., Dame, M.K., Lane, C.L., Smith, C.W., Anderson, D.C., Ward, P.A. *J. Clin. Invest.* 88, 1396–1406.
102. Gundel, R.M., Wegner, C.D., Torcellini, C.A., Clarke, C.C., Haynes, N., Rothlein, R., Smith, C.W., and Letts, L.G. (1991) *J. Clin. Invest.* 88, 1407–1411.
103. Johnston, G.I., Cook, R.G., and McEver, R.P. (1989) *Cell* 56, 1033–1044.
104. Hsu-Lin, S.C., Berman, C.L., Furie, B.C., August, D., and Furie, B. (1984) *J. Biol. Chem.* 259, 9121–9126.
105. Hattori, R., Hamilton, K.K., McEver, R.P., and Sims, P.J. (1989) *J. Biol. Chem.* 264, 9053–9060.
106. Larsen, E., Palabrica, T., Sajer, S., Gilbert, G.E., Wagner, D.D., Furie, B.C., and Furie, B. (1990) *Cell* 63, 467–474.
107. Polley, M.J., Phillips, M.L., Wayner, E., Nudelman, E., Singhal, A.K., Hakomori, S.I., and Paulson, J.C. (1991) *Proc. Natl. Acad. Sci. USA* 88, 6224–6228.

108. McEver, R.P., Beckstead, J.H., Moore, K.L., Marshall-Carlson, L, and Bainton, D.F. (1989) *J. Clin. Invest.* 84, 92–99.
109. Osborn, L., Hession, C., Tizard, R., Vassallo, C., Luhowskyj, S., Chi-Rosso, G., and Lobb, R. (1989) *Cell* 59, 1203–1211.
110. Elices, M.J., Osborn, L., Takada, Y., Crouse, C., Luhowskyj, S., Hemler, M.E., and Lobb, R.R. (1990) *Cell* 60, 577–584.
111. Schwartz, B.R., Wayner, E.A., Carlos, T.M., Ochs, H.D., and Harlan, J.M. (1990) *J. Clin. Invest* 85, 2019–2022.
112. Rice, G.E. and Bevilacqua, M.P. (1989) *Science* 246, 1303–1306.
113. Rice, G.E., Munro, J.M., and Bevilacqua, M.P. (1990) *J. Exp. Med.* 171, 1369–1374.
114. Carlos, T.M., Schwartz, B.R., Kovach, N.L., Yee, E., Rosa, M., Osborn, L., Chi-Rosso, G., Newman, B., Lobb, R., and Harlan, J.M. (1990) *Blood* 76, 965–970.
115. Walsh, G.M., Mermod, J.J., Hartnell, A., Kay A.B., and Wardlaw, A.J. (1991) *J. Immunol.* 146, 3419–3423.
116. Dobrina, A., Menegazzi, R., Carlos, T.M., Nardon, E., Cramer, R., Zacchi, T., Harlan, J.M., and Patriarca, P. (1991) *J. Clin. Invest.* 88, 20–26.
117. Bochner, B.S., Luscinskas, F.H., Gimbrone, M.A., Jr., Newman, W., Sterbinsky, S.A., Derse-Anthony, C.P., Klunk, D., Schleimer, R.P. (1991) *J. Exp. Med.* 173, 1553–1556.
118. Issekutz, T.B. (1991) *FASEB J.* 5, A1335.
119. Geiger, U., Boxer, L.A., Simpson, P.J., Lucchesi, B.R., and Todd, R.F., III (1987) *Blood* 69, 1622–1630.
120. Kehrli, M.E., Jr., Schmalstieg, F.C., Anderson, D.C., Van Der Maaten, M.J., Hughes, B.J., Ackermann, M.R., Wilhelmsen, C.L., Brown, G.B., Stevens, M.G., and Whetstone, C.A. (1990) *Am. J. Vet. Res.* 51, 1826–1836.
121. Harlan, J.M. (1987) *Acta Med. Scand. Suppl.* 715, 123–129.
122. Thomas, J., Gleich, M., Harlan, J., Rice, C., and Winn, R. (1991) *FASEB J.* 5, A509.
123. Bucky, L.P., Vedder, N.B., Hong, C.H.Z., May, J.W., Jr., and Ehrlich, H.P. (1991) *Proc. Amer. Burn Assoc.* 23, 133.
124. Goldman, G., Welbourn, R., Lindsay, T., Hill, J., Shepro, D., and Hechtman, H.B. (1991) *FASEB J.* 5, A1492.
125. Wallace, J.L., Arfors, K.E., and McKnight, G.W. (1991) *Gastroenterology* 100, 878–883.
126. Hutchings, P., Rosen, H., O'Reilly, L., Simpson, E., Gordon, S., and Cook, A. (1990) *Nature* 346, 639–642.
127. Lunn, E.R., Perry, V.H., Brown, M.C., Rosen, H., and Gordon, S. (1989) *Eur. J. Neurosci.* 1, 27–33.
128. Flavin, T., Ivens, K., Rothlein, R., Faanes, R., Clayberger, C., Billingham, M., and Starnes, V.A. (1991) *Transplant. Proc.* 23, 533–534.
129. Wegner, C.D., Wolyniec, W.W., LaPlante, A.M., Marschman, K., Huksk, A., Perigard, C., Rothlein, R., and Letts, L.G. (1991) *Am. Rev. Respir. Dis.* 143, A544.
130. Granger, D.N. (1988) *Am. J. Physiol.* 255, H1269–H1275.
131. Romson, J.L., Hook, B.G., Kunkel, S.L., Abrams, G.D., Schork, A., and Lucchesi, B.R. (1983) *Circulation* 67, 1016–1023.

132. Engler, R.L., Schmid-Schonbein, G.W., Pavelec, R.S. (1983) *Am. J. Pathol.* 111, 98–111.
133. Engler, R.L, Dahlgren, M.D., Morris, D.D., Peterson, M.A., and Schmid-Schonbein, G.W. (1986) *Am. J. Physiol.* 251, H314–H322.
134. Schmid-Schonbein, G.W. and Engler, R.L. (1987) *Am. J. Cardiovasc. Pathol.* (1987) 1, 15–30.
135. Smith, S.M., Holm-Rutili, L., Perry, M.A., Grisham, M.B., Arfors, K.E., Granger, D.N., and Kvietys, P.R. (1987) *Gastroenterology* 93, 466–471.
136. Hernandez, L.A., Grisham, M.B., Twohig, B., Arfors, K.E., Harlan, J.M., and Granger, D.N. (1987) *Am. J. Physiol.* 253, H699–H703.
137. Simpson, P.J., Todd, R.F., Fantone, J.C., Mickelson, J.K., Griffin, J.D., and Lucchesi, B.R. (1988) *J. Clin. Invest.* 81, 624–629.
138. Simpson, P.J., Todd, R.F., III, Mickelson, J.K., Fantone, J.C., Gallagher, K.P., Lee, K.A., Tamura, Y., Cronin, M., and Lucchesi, B.R. (1990) *Circulation* 81, 226–237.
139. Ma X.L., Johnson, G., III, Tsao, P.S., and Lefer, A.M. (1990) *Circ. Res.* 82, III–701.
140. Vedder, N.B., Winn, R.K., Rice, C.L., Chi, E.Y., Arfors, K.E., and Harlan, J.M. (1990) *Proc. Natl. Acad. Sci. USA* 87, 2643–2646.
141. Faist, E., Baue, A.E., Ditmer, H., and Heberger, G. (1983) *J. Trauma* 23, 779–786.
142. Goris, R.J.A., teBoekhorst, T.P.A., Nuytinck, J.K.S., and Gimbrere, J.S.F. (1985) *Arch. Surg.* 120, 1109–1115.
143. Vedder, N.B., Winn, R.K., Rice, C.L., Chi, E.Y., Arfors, K.E., and Harlan, J.M. (1988) *J. Clin. Invest.* 81, 939–944.
144. Vedder, N.B., Fouty, B.W., Winn, R.K., Harlan, J.M., and Rice, C.L. (1989) *Surgery* 106, 509–516.
145. Barroso-Aranda, J., Schmid-Schonbein, G.W., Zweifach, B.W., and Engler, R.L. (1988) *Circ. Res.* 63, 437–447.
146. Mileski, W.J., Winn, R.K., Vedder, N.B., Pohlman, T.H., Harlan, J.M., and Rice, C.L. (1990) *Surgery* 108, 206–212.
147. Welbourn, R., Goldman, G., Hill, J., Lindsay, T., Shepro, D., and Hechtman, H.B. (1991) *FASEB J.* 5, A1492.
148. Bishop, M.J., Kowalski, T.F., Guidotti, S.M., and Harlan, J.M. (1991) *J. Surg. Research.* In press.
149. Carden, D.L., Smith, J.K., and Korthuis, R.J. (1990) *Circ. Res.* 66, 1436–1444.
150. Clark, W.M., Madden, K.P., Rothlein, R., and Zivin, J.A. (1991) *Stroke* 22, 877–883.
151. Mileski, W.J., Harlan, J.M., Heimbach, D., and Rice, C.L. (1990) *Proc. Am. Burn Assoc.* 22, 164.
152. Rosen, H., Gordon, S., and North, R.J. (1989) *J. Exp. Med.* 170, 27–38.
153. Mileski, W.J., Winn, R.K., Harlan, J.M., and Rice, C.L. (1991) *Surgery* 109, 497–501.
154. Sharar, S.R., Winn, R.K., Murry, C.E., Harlan, J.M., and Rice, C.L. (1991) *Surgery* 110, 213–220.

CHAPTER 7

Regulation of Cell Adhesion

Carl G. Figdor, Yvette van Kooyk

Immune surveillance is a major leukocyte function carried out by trafficking of leukocytes throughout the body. This constant recirculation between lymphoid organs and other tissues via lymph and blood implicates adhesion to the endothelial cell layer, migrating underneath the endothelial cells, degrading the basement membrane, and penetrating the underlying tissue. Adhesion and migration of leukocytes is not a random process, but as pointed out in previous chapters of this book, it is orchestrated by a number of specialized adhesion receptors expressed at the cell surface of leukocytes as well as endothelial cells. Different types of lymphocytes home to different lymphoid organs, such as the peripheral and mucosal lymph nodes, by binding to specialized high endothelium in the postcapillary venules,[1] whereas monocytes and granulocytes, but also memory T lymphocytes can specifically adhere to endothelium of inflamed tissues.[2] Important questions that arise from these observations are: What cell surface structures determine lymphocyte homing or adhesion and migration of leukocytes into inflamed tissue, and more importantly, how is this process regulated? Several receptors and counter receptors (adhesion pairs) have now been characterized to mediate binding between leukocytes and endothelial cells (Figure 7-1). Molecular cloning of the genes encoding for these cell surface structures has led to the discovery of distinct groups of structurally related proteins,[3] which have been discussed in detail in the preceding chapters. This chapter focuses on the regulatory mechanisms that determine whether a cell will or will not adhere to endothelium and deals with questions such as: Why does a cell express so many different adhesion receptors at its cell surface? Is there a hierarchy in the usage of different adhesion pathways? How are adhesion receptors engaged in adhesion? Is their communication between different types of adhesion receptors on one and the same cell? This chapter focuses on integrin-mediated (see also Chapter 1) interactions between leukocytes and endothelium, which may be regarded as a model for other cell–cell interactions.

MECHANISMS THAT REGULATE CELL ADHESION

In healthy individuals, binding of circulating leukocytes to flat vascular endothelium is of low avidity, whereas a constant emigration from the blood through the high endothelial cell layer of postcapillary venules into the lymphoid organs can be observed. This correlates with the surveillance function of the immune system. However, the behavior of leukocytes is completely altered once a pathogen (bacteria or viruses) has entered the body and an inflammatory/immune response is initiated. A whole series of events redirects migration of leukocytes to the site of inflammation. Initially, an inflow of granulocytes and, subsequently, monocytes, can be seen. At a later stage, and depending on the type of the response, lymphocytes also adhere to the endothelium, traverse it, and migrate through the underlying subendothelial matrix to the site of inflammation. Both leukocytes and endothelial cells are equipped with several mechanisms to direct and regulate adhesion and migration (Table 7-1). First, upregulation or downregulation of particular adhesion receptors is employed to control adhesive processes. Second, activation and deactivation of adhesion receptors, especially of integrin molecules, play an important role in adhesion, as well as the spatial organization of adhesion receptors at the cell surface. Only very recently, more insight into the communication between different adhesion pairs has been gained and adhesion cascades have been discovered, where adhesion is a result of the subsequent involvement of different adhesion pathways. Differences in glycosylation[4–6] may also contribute to the capacity of adhesion receptors to bind ligand. For example, phorbol myristate acetate (PMA) induces altered glycosylation of the fibronectin receptor very late antigen-5 (VLA-5)[6,7] and of lymphocyte-function antigen-3 (LFA-3),[8] the counter-receptor of CD2.

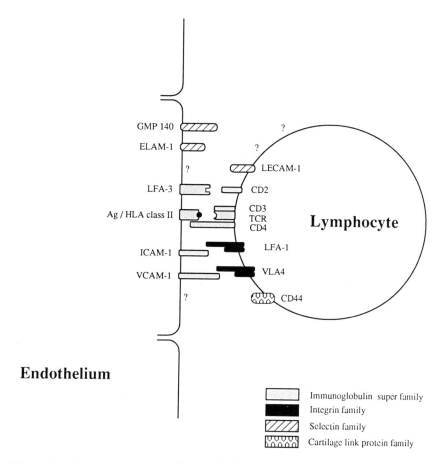

Figure 7-1. Leukocytes use multiple adhesion receptors to bind to endothelium. Antigen (Ag)-specific binding of lymphocytes is mediated by the T cell receptor (TCR) complex, which interacts with Ag presented by MHC class I or II molecules. In addition, several nonspecific adhesion receptors can support lymphocyte adhesion and activation. At least four major groups of adhesion receptors can be distinguished. Note that members of different groups can interact with each other.

METHODS OF REGULATION OF ADHESION RECEPTOR EXPRESSION

Several observations show that the number of adhesion receptors expressed by a cell can be modulated by different factors (Table 7-2). Cell activation by cytokines, chemotactic factors, or infectious agents or products thereof can up- and downregulate cell surface expression. In addition, genetic factors may in part determine expression levels. The expression levels of adhesion receptors do not necessarily correlate with adhesiveness; activation of receptors may be required (Table 7-1).

Table 7-1. Regulation of Cell Adhesion

Regulation at the Level of:	Influenced by:
Number of receptors expressed	Cytokines
	Infectious agents
	Chemotactic compounds
	Genetic factors
Glycosylation of receptors	Cell type specific
Activation of receptors ("inside-out")	Phosphorylation
	Conformational changes
Signalling through receptors ("outside-in")	Second messengers, kinase activity
Receptor organization at the cell surface	Association with other molecules (cytoskeleton)
	Clustering of receptors

Cytokines

During inflammation a large number of different cytokines or bacterial products are produced that have a direct or indirect (by induction of other cytokines: cytokine cascade) effect on the expression of adhesion receptors on endothelial cells as well as leukocytes (see Table 7-2).

LFA-1/ICAM-1/2 Receptor Pair. ICAM-1 (intercellular adhesion molecule-1) is only weakly expressed on mononuclear cells[9] and endothelial cells.[10] Upon stimulation of endothelial cells with inflammatory mediators such as interleukin-1β(IL-1β), interferon-γ (IFN-γ), tumor necrosis factor-α (TNF-α), but also lipopolysaccharide (LPS), surface membrane expression of ICAM-1 increases several-fold.[10-15] Increased expression of ICAM-1 on endothelial cells may lead to enhanced leukocyte–endothelial adhesion.[16,17] ICAM-2, another counter-receptor of LFA-1[18] present on endothelial cells is, in contrast to ICAM-1, constitutively expressed on endothelial cells. Its expression cannot be induced by cytokines.[17,18]

LFA-1 itself (which belongs to the leukocyte-specific integrins: beta 2 group) is less sensitive to cytokine treatment. Although enhanced expression of LFA-1 can be observed in lymphocytes exposed to IL-2, this is limited to

Table 7-2. Regulation of Cell Adhesion by Cytokines and Infectious Agents

Increased/Decreased Adhesion Receptor Expression	Adhesion Molecule	Cell Type		Induced by	Reference
		Leukocyte	Other		
Increase	CD2	Lymphocyte		IL-2	19,34
	LFA-1	Lymphocyte		IL-2, IFN-γ	19,38, 174,175
		Monocyte		IL-4, TGF-β	
	CR3	Monocyte		FMLP, PAF	20–22
		Granulocyte			
	p150, 95	Monocyte		FMLP, PAF	20–22
		Granulocyte			
	ICAM-1	Lymphocyte	Endothelial cell	IL-1β, TNF	10–17
		Monocyte			
		Granulocyte			
	LFA-3	Lymphocyte	Endothelial cell	IL-1β, TNF	19,34
	VCAM-1		Endothelial cell	IL-1β, TNF, IL-4	28,29
	ELAM-1		Endothelial cell	IL-1β, TNF, LPS	44,45,56, 57
	GMP-140		Platelet Endothelial cell	PAF, thrombin	46,47,60
Decrease	LECAM-1	Lymphocyte		LPS, IL-1β	50,51
		Monocyte			
	LFA-1	Lymphocyte		EBV	64–66

a two- to three-fold increase,[19] compared to a five- to 10-fold increase, in ICAM-1 expression. In contrast to LFA-1, a marked increase in expression of CR3 and p150,95, the other members of the beta 2 group of integrins, can be observed[20–22] when granulocytes, and to a lesser extent, monocytes, are stimulated with inflammatory mediators such as chemotactic compounds formyl-methionyl-leucyl-phenylalanine (FMLP) or cytokines (TNF-α, GM-CSF). This increase in expression is rapid, occurring within minutes, which indicates the presence of intracellular stores that contain these receptors. Such storage pools have indeed been localized in granulocytes[23] as specific granules (so-called "tertiary granules"). Interestingly, these granules did not contain LFA-1, indicating that this molecule is directed to the cell surface in a completely different manner. Upon stimulation of the cell, degranulation occurs. However, this has in itself no direct effect on the adhesive behavior of granulocytes,[24,25] indicating that surface expression per se is not sufficient to stimulate adhesion, but that it requires receptor activation (see section entitled Activation of Adhesion Receptors "Inside-Out" Signalling).

VLA-4/VCAM-1 Receptor Pair. In contrast to the leukocyte integrins (beta 2 group), expression of the VLA or beta 1 group of integrins is not restricted to leukocytes.[3,26] Although these integrins predominantly interact with other extracellular matrix components, such as fibronectin, laminin, and collagens,[27] recent observations show that VLA-4 also recognizes a cell surface receptor on endothelial cells: vascular cell adhesion molecule-1 (VCAM-1).[28,29] This molecule, under normal physiologic conditions, is not or is only minimally expressed by endothelial cells, but it is rapidly induced upon stimulation with TNF-α or IL-1β. A murine homologue of VLA-4 (LPAM-2) has also been shown to mediate lymphocyte binding to endothelium of mucosal lymph nodes,[30,30a] but a ligand has not yet been determined. Furthermore, VLA-4 has been reported to mediate cytotoxic T-cell activity[31,32] of particular human T lymphocytes, although this is probably not mediated by VCAM-1, as leukocytes, which were used as target cells, do not express VCAM-1. In addition to VCAM-1, VLA-4 expresses a fibronectin binding site,[26,33] different from the site that is recognized by VLA-5, the classical fibronectin receptor. It can not, therefore, be excluded that VLA-4 contributes to cytotoxic activity by binding to fibronectin. Fibronectin may bind to target cells that express VLA-5 and, thus, form a bridge between VLA-4 and VLA-5 by binding to different sites on the fibronectin molecule.[26]

Although the ligand of VLA-4 is readily upregulated by cytokines, VLA-4 expression, but also VLA-5 and VLA-6 expression, is only marginally affected by cytokines. This parallels observations made for the LFA-1/ICAM-1 as well as the CD2/LFA-3 (see below) adhesion pathway, suggesting that increase in expression by cytokines is predominantly found at the level of the ligand, and that, apart from a modest increase in expression during lymphocyte maturation/differentiation,[19,34] the number of receptor molecules remains approximately constant.

CD2/LFA-3 Receptor Pair. A strong increase in LFA-3 expression can be observed after treatment of endothelial cells or lymphocytes with cytokines such as IFN-γ on IL-2, respectively.[19,34] It has been extensively reported in the literature that the CD2/LFA-3 pathway is implicated in recognition of target cells by cytotoxic lymphocytes,[35] but a clear role in adhesion of lymphocytes to endothelial cells has not been documented. Only recently, Hughes et al.[36] reported that IL-2 production induced by binding of lymphocytes to endothelial cells is completely dependent on the CD2/LFA-3 adhesion pair, suggesting that this adhesion pathway is indeed of relevance in lymphocyte–endothelial cell interactions. Similar to LFA-1, the increase in expression of CD2 on lymphocytes induced by cytokines is modest compared to the huge increase in expression of LFA-3 (from virtually absent to high levels of expression).[37] An exception may be transforming growth factor-β (TGF-β), which is able to increase the expression of several beta 1, beta 2 and beta 3 integrins.[38,39]

Selectins. A novel class of structurally-related adhesion molecules has been termed *selectins*. This group consists of three members: the so-called *lymphocyte homing receptor* (also termed *MEL-14, LAM-1, LECAM-1, or LECCAM-1*);[1,40–44] *ELAM-1*;[45] and *GMP-140* (also called *PADGEM* or *CD62*).[46,47] A common feature of these molecules is that they contain a lectin-binding domain, and recent findings indeed confirm that the potential counterstructures recognized by the selectins are carbohydrate structures.[1,41,42,44,45,48,49]

LECAM-1 is expressed by leukocytes and, although it has a function in homing of lymphocytes to peripheral lymph nodes, its expression on granulocytes predicts a more general function. The rapid downregulation observed upon stimulation of granulocytes with phorbol ester (PMA) or IL-1β[50,51] suggests that it plays a role early in the adhesive process. When lymphocytes are cultured in the presence of IL-2 they also lose LECAM-1 expression (unpublished results). In addition, all *in vitro* cultured T-cell clones (CD4 or CD8) as well as the NK clones are LECAM-1-negative (unpublished observations). This may be entirely due to the fact that *in vitro*, constant stimulation (cytokine, antigen) is required to promote lymphocyte proliferation, and it may be entirely different from the *in vivo* situation. Indeed, Tedder et al.[52] demonstrates that memory cells of the helper phenotype are LECAM-1 positive, which supports the notion that LECAM-1 has an important function in immune surveillance.

Despite the fact that ELAM-1, another member of the selectin family, can be expressed by endothelial cells whereas LECAM-1 is expressed by leukocytes, there is no direct evidence that both molecules bind to each other.[53] Most recent observations indicate that certain subpopulations of lymphocytes may also use ELAM-1 to extravasate.[54,55] ELAM-1 is normally absent on endothelial cells but becomes rapidly expressed when endothelial cells are stimulated with TNF-α, IL-1β, or LPS.[12,44,56,57] It peaks at four to six hours, after which expression diminishes to baseline levels after 24 hours. This transient expression of ELAM-1 is in vast contrast to the prolonged expression of ICAM-1, LFA-3, and VCAM-1 on endothelial cells after cytokine treatment. These differences in the kinetics of the expression of these adhesion molecules are a clear example of a mechanism that can be employed by endothelial cells to recruit different types of cells from the circulation.

CD62 (also known as PADGEM or GMP-140) is yet another example of an inducible adhesion molecule on endothelial cells and platelets. This protein, which is extensively discussed in Chapter 2, is released from the Weibel-Palade bodies[58,59] in endothelial cells upon stimulation with thrombin.[60] Increasing evidence suggests that it may recognize CD15[61] and mediate binding of granulocytes and monocytes, but not of lymphocytes, to endothelial cells in a CD18-independent manner.[59,62]

The selectins fulfill their function probably very early in the adhesion process. They are involved in a process termed "rolling of leukocytes" (rolling receptors) along the endothelial cells in the capillaries,[63] thereby providing the cell with a mechanism to rapidly adhere to and penetrate the vessel wall.

Infectious Agents can Modulate Adhesion Receptor Expression

Downregulation of LFA-1 and ICAM-1, but also of LFA-3, has been observed in certain pathologic situations, such as Burkitt lymphoma.[64-66] Preferentially Epstein-Barr virus (EBV)-negative Burkitt lymphoma patients seem to express lower levels of adhesion receptors, though there is no absolute correlation.[67] The role of EBV in adhesion molecule expression in Burkitt lymphoma is not completely understood, although it has been demonstrated that viral infection can enhance LFA-1 mRNA levels and surface expression stimulated by the latent membrane protein (LMP) gene of the EBV virus.[68] Similarly, bacterial infections (bacillus Calmette-Guérin) have also been reported to enhance LFA-1 expression.[69]

Genetic Factors

In addition to the effect of cytokines or infectious agents, genetic factors also seem to play a role. It was previously recognized that freshly-isolated peripheral blood lymphocytes contain cells that express high-end low levels of LFA-1. Lymphocytes expressing high levels of LFA-1 were assigned to the so-called "memory phenotype," lymphocytes that have previously been exposed to antigen. On the other hand, the LFA-1 low lymphocytes are thought to represent the virgin lymphocytes, newly immigrated from the thymus.[34] However, an intriguing observation made by Bender et al. demonstrates that, in addition to cell maturation and differentiation, genetic factors also determine the level of LFA-1 expression in normal individuals.[70] They investigated the levels of LFA-1 expressed by peripheral lymphocytes from 20 pairs of identical twins and observed remarkably similar fluorescence activated cell sorter (FACS) profiles between identical twins but large differences in expression when different pairs of twins were compared. Considering the fact that it is highly unlikely that the immune status between twins is always similar (determined by environmental factors), one must conclude that the level of LFA-1 expression is, at least partially, determined by genetic factors.

ACTIVATION OF ADHESION RECEPTORS ("INSIDE-OUT" SIGNALLING)

Several observations, predominantly made during studies of integrin molecules, indicate that a second mechanism to control cell adhesion is modulation of the affinity of the receptor for its ligand. Resting leukocytes or platelets do not adhere spontaneously, but a variety of stimuli can induce beta 1- (VLA-4, VLA-5, VLA-6), beta 2- (LFA-1 , CR3), and beta 3- integrin (IIb/IIIa)-mediated cell–cell interactions (Table 7-3). Exposure of lymphocytes, myeloid cells, or platelets to phorbol ester (PMA) strongly induces cell aggregation.[17,71-75] Similarly, FMLP can stimulate CR3-mediated adhesion of granulocytes to endothelial cells,[24,25,76] and activation of platelets by thrombin or adenosine diphosphate (ADP) causes IIb/IIIa-mediated aggregation.[77-79] A prominent characteristic in all of these observations is that adhesion is induced without an apparent increase in receptor expression. This suggests that changes in affinity of the receptor for its ligand, or changes in the organization of the adhesion receptors at the cell surface, are likely to cause enhanced cell adhesion. Second messengers play a pivotal role in integrin activation, although at present the precise intracellular circuits that regulate integrin-mediated cell adhesion are not yet completely understood.

Antigen-Specific and Nonspecific Activation of LFA-1

Recently, it has been found that, except for PMA, monoclonal antibodies directed against the cell surface molecules expressed by T cells (such as CD2 and CD3) can stimulate cell[80,81] aggregation (Figure 7-2) or binding to endothelium.[82] These observations indicate that surface molecules such as

Table 7-3. Activation of Adhesion Receptors: "Inside-Out" Signalling

Receptor	Expressed by	Activation Induced by
LFA-1	Ly, Mo, Gr*	PMA, αMHC class II, αCD2, αCD3, αCD40, αCD43, αCD44
CR3	Ly, Mo, Gr	PMA, ADP, ATP, FMLP, PAF, IL8, αCD14, αCD15
VLA-4	Ly, Mo	PMA, αCD2, αCD3
VLA-5	Mo, Ly	PMA, αCD2, αCD3
VLA-6	Mo, Ly	PMA, αCD2, αCD3
IIb/IIIa	Plt*	PMA, ADP, PAF, thrombin

*Mo = monocyte; Ly = lymphocyte; Gr = granulocyte; Plt = platelet.

CD2 and CD3 can activate LFA-1 via intracellular signalling pathways and that activation of LFA-1 leads to enhanced ligand binding, which implies the existence of an inactive and an active form of LFA-1. A strong argument in favor of the existence of these two states of LFA-1 is that if only one form of LFA-1 existed, spontaneous aggregation of peripheral blood leukocytes would cause injury of the microvascular network, induce microembolization, and may thereby compromise normal leukocyte circulation. Additional evidence for this hypothesis comes from the observation that freshly-isolated resting leukocytes do not tend to aggregate to each other, although they express significant levels of LFA-1 and ICAM-1. Similarly, cloned cytotoxic T lymphocytes (CTL) and natural killer (NK) cells, which express extremely high levels of LFA-1 and ICAM-1, do not aggregate offhand.[80] Only upon stimulation with antigen, PMA, or via CD2 or CD3 by mAbs, rapid cell aggregation (<20 minutes) of cytolytic T lymphocytes (CTL) can be observed. This cell clustering is LFA-1-dependent, as it is abrogated by anti-LFA-1 antibodies. In addition, surface expression of LFA-1 and ICAM-1 does not change during activation of T-cell or NK-cell clones,[80] indicating that cell aggregation is not caused just by augmented surface expression. Other reports show that antibodies directed against CD43,[83] CD44,[84] CD40 on B cells,[85] CD14 on monocytes,[86] CD15,[87] and major class II histocompatibility complex antigens (MHC)[88] induce LFA-1-mediated adhesion (see Table 7-3). These observations support the hypothesis that activation of LFA-1 can be induced through different surface receptors, indicating that the LFA-1/ICAM-1 pathway is a common adhesion pathway that may be used by leukocytes under quite different physiologic conditions.

Interestingly, it has been observed that, depending on the stimulus used, activation of LFA-1 was either transient (20 minutes) or sustained[80] for a prolonged period of time (≥two hours). These results indicate that different and complex signalling pathways may be employed to regulate activation of LFA-1 to ensure the required stable or transient interactions in the myriad of functions mediated by this molecule. A good example to demonstrate the mode of action of LFA-1 is the cytotoxic T cell (CTL)–target cell interaction, the effector phase of an immune response. When a CTL encounters a target cell, initial cell–cell contact is established through LFA-1–ICAM interaction. Cell binding is nonspecific and of low avidity as it does not implicate antigen recognition by the T-cell receptor. Nevertheless, target-cell binding is strong enough to facilitate recognition of antigen by the T-cell receptor/CD3 complex. If no antigen is recognized, the CTL detaches from the target cell. If, however, antigen is recognized by the T-cell receptor, CD3 creates a signal leading to activation of LFA-1. Activated LFA-1 then mediates high-avidity binding between CTL and target cell, thereby strengthening the adhesion between both cells.[35] This facilitates the formation of intercellular clefts,[89] enabling efficient delivery of cytotoxic molecules into the target cell. In

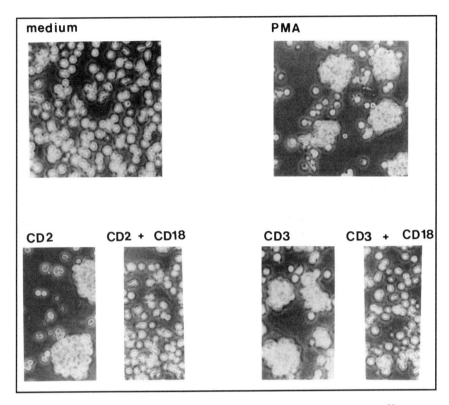

Figure 7-2. PMA, crosslinking of CD2, or of CD3 by appropriate antibodies[80] induces strong aggregation of cloned T lymphocytes. This aggregation is LFA-1-dependent, as anti-CD18 antibodies abrogate aggregation.

addition, activated LFA-1 may interact with cytoskeletal elements, thereby directing the migration of cytotoxic granules. Modulation of the TCR/CD3 complex from the cell surface after peptide MHC binding, possibly by phosphorylation of the CD3 gamma component[90–92] may abrogate signalling, thus acting as a negative feedback signal, reversing LFA-1 into its inactive state and providing the CTL of a mechanism to detach from a target cell.[93] In the induction phase of an immune response, when antigen is presented by a MHC molecule to the T-cell receptor, LFA-1, both on the antigen-presenting cell (B cell, macrophage) and on the responder cell, may be engaged in cell–cell contact, as CD3 may activate LFA-1 on the T cell and MHC class II may stimulate LFA-1 on the antigen-presenting cell.[88] Activation and deactivation of LFA-1 provides leukocytes with a general mechanism to attach and detach to cells and to migrate throughout the body. Because the ligands of LFA-1 are widely spread, lymphocytes can migrate to essentially every organ in the body.

Activation of LFA-1 Compared to Other Integrins

As summarized in Table 7-3, a still-increasing list of soluble agents (cytokines, chemotactic factors, coagulation factors) and cell surface receptors have been defined to induce integrin-mediated adhesion.

FMLP, IL-8, or the nucleotides ATP or ADP stimulate CR3-mediated adhesion to endothelium or homotypic aggregation of granulocytes in a transient manner, similar to LFA-1.[24,25,94–97] Also, members of the beta 1 family, VLA-4, VLA-5, and VLA-6, expressed by CD4+ cells can become activated through PMA, CD2, or CD3 and show sustained or transient binding to their respective ligands (fibronectin and laminin).[98,99] Furthermore, it has been shown that the platelet antigen IIb/IIIa, a member of the beta 3 group of the integrins, avidly binds to arginine-glycine-aspartate (RGD)-containing peptides,[100–103] but requires activation of the receptor in order to bind fibrinogen.[77,104] These observations show that several members of the integrin family require activation to manifest full competence to bind ligands. Modulation of the affinity for ligand thus seems a common way to regulate integrin receptor function.

Intracellular Messengers and Affinity Modulation of Integrin Receptors

PMA, a potent activator of protein kinase C (PKC), may induce phosphorylation of a variety of surface proteins,[105] including LFA-1. Increased phosphorylation of the beta 2 chain on serine residues has indeed been reported upon stimulation with PMA or with FMLP, whereas the α chains of the beta 2 group of integrins seem to be constitutively phosphorylated.[106,107] The intracellular domain of the beta 2 chain contains several potential phosphorylation sites on serine/threonine residues.[3,107] Deletion of the cytoplasmic domain, including these phosphorylation sites, leads to loss of function of LFA-1.[108] This does not, however, prove that phosphorylation of the beta 2 chain is essential but may point to important interactions with intracellular proteins (cytoskeleton). Unpublished observations show that phosphorylation of beta 2 induced by PMA is weak compared to phosphorylation of a number of other surface proteins. Only amino acid substitution experiments will address the role of phosphorylation of beta 2 in modulating the affinity of the LFA-1 molecule. On the other hand, it cannot be excluded that phosphorylation of cytoskeletal components (see next section), which are in close proximity or possibly linked to integrins, may affect the affinity status of the receptor. PMA and FMLP use different signalling pathways to activate CR3.[109] PMA-induced adhesion is sustained and PKC-dependent, as it can be blocked completely by staurosporine, a PKC inhibitor. Conversely, FMLP-induced adhesion is transient and not inhibited by staurosporine, which seems to enhance and prolong adhesion.[109]

Whereas PMA can directly activate PKC, activation through CD2, CD3, but also CD43[83] is thought to stimulate inositol phospholipid metabolism (Figure 7-3), thereby giving rise to activation of PKC and a rise in $[Ca^{2+}]_i$ levels.[110,111] Recent reports suggest that activation of tyrosine kinases would precede phospho-inositol turnover.[112] Although the intracellular domain of the beta 2 chain contains one tyrosine residue, phosphorylation on tyrosine has not been documented (unpublished data). Several reports show a direct correlation between a rise in $[Ca^{2+}]_i$ levels and the high affinity status of the leukocyte integrin receptors,[87,94] indicating that $[Ca^{2+}]_i$ plays an important role in activation of LFA-1. In addition, anti-CD44 antibodies have been shown to enhance LFA-1-mediated binding.[84] However, these antibodies were not able to stimulate PKC or to mobilize $[Ca^{2+}]_i$ from intracellular pools (unpublished results), suggesting that another signal transduction pathway is employed. The notion that CD44 has a signalling function is underscored by experiments showing that anti-CD44 antibodies are costimulatory, together with anti-CD3, in inducing lymphocyte proliferation.[113]

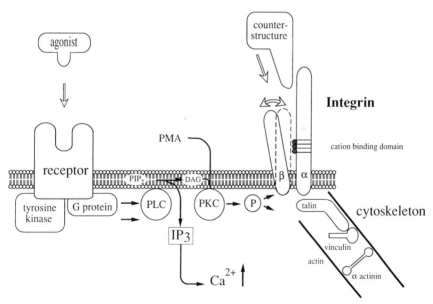

Figure 7-3. Integrin receptors can be activated by triggering a variety of other cell surface receptors on leukocytes (see Table 7-3). Binding of ligand (or crosslinking of receptor by antibodies) induces G protein/tyrosine kinase activity resulting in phospholipase C (PLC)-mediated phosphoinositol phosphate (PIP) hydrolysis. Subsequent release of diacylglycerol (DAG) results in protein kinase C (PKC) activation and IP$_3$-mediated release of $[Ca^{2+}]_i$. The exact mechanism that causes integrin receptor activation is not known, but association with the cytoskeleton, possibly by phosphorylation of cytoskeletal components or of the receptor, has been suggested. Activation of integrin receptors results in enhanced binding to its counter-structure.

Activation Epitopes, the Role of Cations and Receptor Conformation

Much has been learned on the conditions required for integrin–ligand interaction. Integrin-mediated cell adhesion is a temperature- and energy-dependent process that requires an intact cytoskeleton and the presence of divalent cations, notably Mg^{2+} and/or Ca^{2+}.[114] In most cases, Ca^{2+} or Mg^{2+} can be substituted by other divalent cations. Interestingly, Mn^{2+} causes spontaneous activation of LFA-1 resulting in cell aggregation (unpublished data). The underlying mechanism and its physiologic relevance is unknown, but the observation underscores the importance of divalent cations in activation of integrins. Cloning of the genes encoding the integrin molecules revealed the presence of three or four potential cation binding sites on the various α chains.[3,27] A number of observations underline the important role of these metal ions in integrin-receptor activation.

An anti-LFA-1 antibody has been described,[115,116] designated NKI-L16 (called *L16*), that in contrast to other anti-LFA-1 antibodies, stimulates cell adhesion rather than inhibiting LFA-1-dependent cell–cell interactions. L16-induced cell aggregation does not implicate "outside-in" signalling, which will be discussed later, but is merely thought to act by modulating the conformation of LFA-1, so that the affinity for ligand binding is greatly enhanced.[116] Time course studies measuring cell aggregation induced by L16 or by F(ab)′ fragments thereof showed a kinetics strikingly similar to that observed when cells were stimulated with PMA.[115] This observation led to the suggestion that stimulation of cells with PMA may also result in a conformational change of LFA-1, thus increasing the affinity for its ligands.

Recently, it was found that the epitope on LFA-1 recognized by the L16 antibody is Ca^{2+}-dependent. Treatment of cloned T cells, which express L16 abundantly, with a metal chelating agent (EDTA or EGTA) results in a loss of this epitope and, more importantly, lose the capacity to bind to other cells in a LFA-1-dependent manner.[116] Moreover, loss of the L16 epitope is not associated with a reduction of other epitopes of LFA-1, showing that the molecule is still present at the cell surface.[116] In addition, resting peripheral blood leukocytes (PBL) express LFA-1 on their cell surface but they virtually lack expression of the L16 epitope in comparison with cloned T lymphocytes.[115,116] It was observed that the ability of lymphocytes to aggregate upon stimulation directly correlates with expression of this L16 epitope on LFA-1[116] (unpublished data), suggesting that expression of the L16 epitope is a prerequisite for LFA-1-dependent adhesion. This notion is supported by the finding that stimulation of resting lymphocytes with IL-2 or PMA (3–12 hours) results in a significant increase in expression of the L16 epitope and restored their capacity to aggregate. This is not due to an increased number of LFA-1 molecules expressed on the cell membrane, as it is not accompanied by a concomitant rise in expression of LFA-1 epitopes other than L16.[116]

These observations show that the majority of LFA-1 molecules expressed by resting lymphocytes lack the L16 epitope, but that this epitope becomes readily expressed upon stimulation of the cells (Figure 7-4). T-cell or NK-clones express high levels of LFA-1 molecules, all of which expose the L16 epitope. However, expression of the L16 epitope is not sufficient to induce cell adhesion, as these cells do not spontaneously aggregate.[80] Expression of the L16 epitope is a prerequisite for LFA-1-mediated cell binding; it potentiates LFA-1 to become engaged in cell adhesion, but expression of the epitope in itself is apparently not sufficient to induce cell adhesion. It requires a strong stimulatory signal (antigen, PMA, anti-CD3, etc.), which then leads to a third form of LFA-1, activated LFA-1 (see Figure 7-4), which enables stable cell binding. The L16 epitope does not bind to the ligand binding domain, as addition of the antibody in itself induces LFA-1-dependent lymphocyte aggregation or binding of lymphocytes to endothelial cells[80] (unpublished data). This notion is supported by the observation that L16-induced adhesion is completely blocked by other anti- LFA-1α or -β antibodies.[80,116] Because F(ab)' fragments of L16 were also capable of inducing adhesion, it can be excluded that crosslinking of receptors or Fc receptor-mediated phenomena are involved. The observation that neither PKC inhibitors (staurosporin, AMG) nor sodium azide were capable of inhibiting L16-induced adhesion, but unequivocally inhibited PMA-stimulated adhesion, suggests that L16 does not induce signalling into the cell.[116] Together, these results suggest that L16 induces a change in the tertiary structure of LFA-1, which may result in modulation of the ligand binding affinity.

What is the benefit of expressing these distinct forms of LFA-1? As discussed above, clear advantages can be envisaged if LFA-1–ligand interactions can be switched on and off, thereby creating a mechanism to regulate leukocyte adhesion and de-adhesion. But this does not necessitate the existence of an L16+ and L16– form of LFA-1. The answer to this question may be found in the maturation/activation state of a leukocyte. This is best illustrated by the following example. One of the early phases of an immune/inflammatory response is characterized by the recruitment of leukocytes from the peripheral blood pool. Lymphocytes adhere to endothelial cells and migrate into the underlying tissues, a process that is regulated by various adhesion pathways, including LFA-1–ICAM-1 interactions. This type of LFA-1-mediated adhesion is regulated at different levels. Cytokine production will locally raise the expression of ICAM-1[17] on the endothelial cells, thereby facilitating cell adhesion in general. However, it is preferable that, instead of random binding of lymphocytes, only those cells that can positively contribute to an immune response bind to the endothelium. Therefore, immature resting lymphocytes may not be capable of binding unless they express the L16 epitope. Previously activated ("memory") T cells, exposing the L16 epitope, can rapidly bind upon activation of LFA-1[117] by,

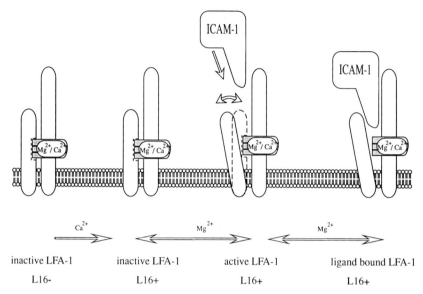

Figure 7-4. Four distinct forms of LFA-1 can be distinguished: inactive LFA-1, which lacks the L16 epitope; intermediate form of LFA-1, which expresses the L16 epitope, but does not mediate high affinity ligand binding; activated LFA-1 (L16$^+$), which mediates high affinity ligand binding; and the ligand bound form.

for instance, peptide MHC presented by endothelial cells. In addition, release of cytokines may result in activation/maturation of resting cells, resulting in exposure of the L16 epitope, thereby gaining the capacity to adhere.

Recently, another antibody, termed *24*,[118] has been described that recognizes a cation-dependent epitope on the leukocyte integrins. This antibody recognizes an epitope completely different from the one recognized by L16 (Table 7-4). First, 24 recognizes an epitope expressed by all three α subunits of the leukocyte integrins, whereas L16 only recognizes LFA-1α. Second, binding of 24 on intact cells is Mg^{2+}-dependent (this requirement is lost when both subunits are dissociated in detergent solution,[119] whereas L16 requires the presence of Ca^{2+}. Third, it has been suggested that expression of the 24 epitope is indicative of the functional competent state of the receptor, whereas L16 recognizes a state intermediate between an inactive and a fully activated receptor. Fourth, antibody 24 recognizes a temperature-dependent epitope (not expressed at 4°C); the L16 epitope is expressed both at 4°C and at 37°C. Fifth, L16 epitope expression is acquired during cell maturation or cell differentiation,[93,116] which has not been reported for the epitope recognized by antibody 24. Sixth, L16 induces cell adhesion, whereas such a characteristic has not (yet) been documented for antibody 24. A seventh distinction between 24 and L16 is the observation that the L16 epitope, once acquired, is never lost, whereas the 24 epitope may only be expressed when the receptor is in its active conformation.

Table 7-4. Comparison of Activation Epitopes on Integrins

Reference	Dransfield (118)	v. Kooyk (116)	Gulino (129)	O'Toole (104)	O'Toole (104)	Shattil (79)	Frelinger (103,126) LIBS1	PMI 1	PMI 2	Kouns (127)
Antibody	24	L16	D33C	P41	62	PAC-1	LIBS1	PMI 1	PMI 2	D3GP
Antigen	LFA-1α CR3 α, p150,95 α,	LFA-1a	IIb	IIIa	IIIa	IIb/IIIa	IIb	IIb	IIb	IIIa
Ligand	ICAM-1	ICAM-1	FGN	FGN	FGN	FGN	FGN	FGN	FGN	FGN
Ab Induces Activated State of Receptor By										
IgG	—	+	+	+	+	—	—	—	—	+
Fab'	—	+	+	+	+	—	—	—	—	+
Maintains Active State of the Receptor	?	+	+	—	—	?	—	—	—	?
Epitope:										
Cation dep. expression	Mg^{2+}	Ca^{2+}	Ca^{2+}	—	—	Ca^{2+}	—	—	—	?
Energy dep. expression	+	—	—	—	—	—	—	—	—	?
Cation binding domain	?	?	+	—	—	?	—	—	—	?
Ligand binding domain	?	—	—	—	—	—	—	—	—	?
LIBS	?	—	—	—	—	—	+	+	+	—

Mg^{2+} ions and at suboptimal concentrations also Ca^{2+} ions play an important role in LFA-1-mediated binding,[17] indicating that Mg^{2+} plays a regulatory role in activating and deactivating the receptor. The inactive state of the receptor might be characterized with the incapability to bind cations and that expression of the 24 epitope indicates the Mg^{2+} bound and active status of the receptor.[119] An alternative hypothesis may be that when LFA-1 expresses the L16 epitope, the molecule constantly "flip-flops" from an inactive into an active state, and that the addition of L16 locks LFA-1 in the active state, thus explaining the induction of cell adhesion by L16. Because the L16 antibody is able to activate LFA-1, it would be interesting to determine if it induces the epitope recognized by antibody 24. Together, both antibodies indicate an important role of cations in modulation of the binding activity of LFA-1. Both L16 and 24 may be indicative for binding of Ca^{2+} or Mg^{2+} to LFA-1, respectively, and may determine a specific but distinct functional status of the receptor. The leukocyte integrins express three cation binding domains.[120] This allows the possibility that occupancy by either Ca^{2+} or Mg^{2+} alone or a combination of both may have dramatic consequences for the functional status of the receptor. Integrins that have three potential metal binding repeats exhibit primarily Mg^{2+}-dependent function.[3,121,122] Stimulation of cell aggregation with PMA absolutely requires the presence of Mg^{2+} and exposure of cloned T lymphocytes to EDTA completely abrogates the capacity of these cells to aggregate. However reconstitution of the cells with Mg^{2+} but not Ca^{2+} causes dramatic spontaneous aggregation that may be due to occupancy of all three metal binding sites by Mg^{2+} (unpublished data), supporting the hypothesis that Mg^{2+} may induce an activated state of the receptor. On the other hand, it was found that in the absence of Mg^{2+}, and possibly also in the absence of the epitope recognized by antibody 24, L16 is still capable to induce LFA-1-mediated cell adhesion, demonstrating that LFA-1-mediated adhesion can also occur in the complete absence of Mg^{2+}. In addition, it was observed that the addition of Mn^{2+} to the medium immediately results in induction of LFA-1-dependent adhesion (unpublished data). This may indicate that Mn^{2+} can substitute Mg^{2+} or Ca^{2+} and is able to activate LFA-1. However, it is not clear whether Mn^{2+} really binds to LFA-1 or stimulates cell aggregation by another unknown mechanism.

Experimental data[80,116] and preliminary results[119] suggest that the epitopes recognized by these two antibodies are not within the ligand binding domain, indicating that (and in contrast to findings with the beta 3 integrins)[123,124] the cation binding domains of leukocyte integrins are not directly involved in ligand binding. Binding of the RGD peptide to the vitronectin receptor requires both the α and the β subunit, and one of the two RGD binding sites resides in the metal binding domains.[123] Also, the cation binding repeats of the α chain of the platelet antigen IIb/IIIa are implicated in binding to fibrinogen,[124] and this group demonstrated recently that a short peptide homologous to the carboxyterminus of the γ chain of fibrinogen is capable to bind to gp IIb. The 11-residue peptide from this region that included the calcium binding motive was efficiently capable to inhibit platelet aggregation and fibrinogen binding to isolated IIb/IIIa and interacted directly with fibrinogen, thus emphasizing the important role of cation-binding sites in receptor function.[125] Binding of fibrinogen as a result of stimulation of platelets with PMA, thrombin, ADP, or epinephrine, induces the expression of a new epitope (termed *LIBS,* ligand induced binding site)[126] recognized by antibody PMI-1.[103] This antibody recognizes a clearly different type of epitope when compared with those recognized by L16 and 24. It preferentially recognizes the ligand bound form of the receptor,[126] as opposed to L16 and 24, which both recognize the unoccupied form. Recently, an elegant study of O'Toole et al.[104] described two antibodies (P41 and Ab-62) with characteristics remarkably similar to those of L16 (see Table 7-4). IIb/IIIa quickly binds substrate bound fibrinogen or RGD peptide but requires activation to bind soluble fibrinogen.[77] Activation is associated with the expression of a new epitope recognized by the PAC-1 antibody.[79] They studied the capacity of stable IIb/IIIa transfectants to bind soluble fibrinogen. They found that the recombinant IIb/IIIa receptor required activation in order to avidly bind fibrinogen, similar to the native receptor in platelets, showing that the lack of high-affinity binding of this ligand by resting platelets is not due to a unique property of the platelet surface micro-environment. They observed that both aforementioned antibodies or PMA were equally capable to induce fibrinogen binding and the PAC-1 epitope. In addition, F(ab)' fragments of both antibodies had similar activating properties. Induction of ligand binding by the antibodies, but not by PMA, was independent of signal transduction events. Binding of fibrinogen could still be demonstrated after fixation of the platelets with paraformaldehyde. Similar findings were made by Kouns et al., who described different antibodies that activated IIb/IIIa without changes in intracellular Ca^{2+} levels, protein phosphorylation, or phospholipid metabolism.[127] O'Toole et al. showed that even the presence of a cell membrane micro-environment (isolated receptor) was not required to induce the activated state of the receptor.[104] These data indicate that activation of IIb/IIa is an inherent property of the receptor. This latter finding is in contrast with observations made so far with purified LFA-1, which always seems to be in an active conformation.[3] Similarly, when LFA-1 is transiently expressed by COS cells, it avidly binds to ICAM-1 and cannot be activated by agonists such as PMA,[128] suggesting that it is expressed in its active form. The findings of O'Toole are one of the first to report that activation is an intrinsic property of the receptor itself, and strongly suggest that this is caused by an alteration in the conformation of the receptor, allowing high-affinity ligand binding.

In addition to antigen beta 3 subunit antibodies that activate platelet aggregation and binding to soluble fibrinogen, a similar antibody directed to the α subunit of the IIb/IIIa receptor has also been described.[129] This antibody, D33C, has even more resemblance with the L16 antibody than the two mentioned above (see Table 7-4). Similar to L16, it reacts with the α chain of an integrin molecule and it recognizes a Ca^{2+}-dependent epitope. Gulino et al. found that this antibody recognizes a putative calcium binding site on IIb, as corresponding synthetic peptides could inhibit antibody binding.[129] It is tempting to speculate that L16 recognizes a similar site located at, or in close proximity of, the highly conserved cation binding domains of LFA-1, and it will be interesting to see if synthetic peptides that correspond to the potential antigenic sequences within the structure of LFA-1 inhibit binding of the L16 activity.

Recently, adhesion-inducing antibodies directed against the beta 1 group have also been identified. Anti-VLA-4 α chain antibodies were found to induce aggregation of leukocytes.[130,131] These antibodies, however, do not recognize a cation-dependent epitope. One of those, L25, has also been reported to inhibit cytotoxic activity,[132] which may indicate that the binding site of L25 is in close proximity of a ligand binding site involved in mediating CTL–target cell interactions. Cell aggregation induced by anti-VLA-4 antibodies is independent from that mediated by the leukocyte integrins, but metabolic energy as well as an intact cytoskeleton are required. It cannot be excluded that, similar to L16, or D33C, a conformational change stimulated by these antibodies induces an activated state of the VLA-4 receptor. A ligand that mediates these interactions has not yet been defined, but VCAM-1 does not seem to be a likely candidate, because it is not expressed on at least part of the cells that were used in these studies. Finally, an antibody directed against the beta 1 subunit (TASK) has been demonstrated to promote adhesion of retinal neurons to laminin and collagen types I and IV, while at the same time it is able to inhibit binding to vitronectin.[133]

SIGNALLING THROUGH ADHESION RECEPTORS ("OUTSIDE-IN" SIGNALLING)

In the previous sections, it was seen that adhesion can be modulated by receptor expression or by altering the affinity state of an adhesion receptor. This section summarizes the increasing amount of literature that indicates that the adhesion receptors themselves (particularly integrin receptors) are also capable of transmitting transient signals into the interior of a cell ("outside-in" signalling). Signalling through adhesion receptors may inform a cell of its micro-environment and may influence cell maturation/differentiation or activation. Clustering of integrins at the cell surface by their ligand on an adjacent cell may generate signals that maintain or extend the active state of the receptor. In most cases reported, evidence is still indirect, as the contribution of the signal provided by an adhesion receptor in itself is seldomly sufficient to trigger cell function. This is in agreement with the hypothesis that integrins have an accessory role in immune and inflammatory responses and are directed by other cell surface molecules such as a T-cell receptor or a receptor for chemotactic peptides. Intracellular messengers that mediate signal transduction are often unknown but may be different from the known signalling pathways (such as phosphoinositol metabolism, $[Ca^{2+}]_i$ levels, tyrosine kinase activity, etc). However, in some reports, triggering of the receptor resulted in a rise in intracellular calcium levels or phosphoinositide hydrolysis.[134,135] Pardi et al.[134] showed that crosslinking of LFA-1 by anti-α but not by anti-β antibodies causes phosphoinositide hydrolysis and a rise in $[Ca^{2+}]_i$. In addition, Pircher et al. showed that antibodies against LFA-1 were able to induce proliferation and cytokine production.[136] Richter et al.[137] demonstrated that the secretion of TNF-α by neutrophils was associated with CR3-mediated oscillations in $[Ca^{2+}]_i$, providing further evidence that Ca^{2+} signals generated by adherence through integrins may alter the functional properties of the cell. Several reports indicate that anti-LFA-1α antibodies but not anti-LFA-1β antibodies can, together with anti-CD3 antibodies, enhance lymphocyte proliferation, suggesting that LFA-1 may act as a signalling molecule.[138] Other reports support the observation that anti-LFA-1 antibodies are costimulatory with anti-CD3 but could not make a distinction between anti-α or -β subunit antibodies.[135] Similarly, binding of lymphocytes to immobilized affinity purified ICAM-1 and anti-CD3 antibodies induced cell proliferation,[99] supporting the notion that ICAM-1 can provide a costimulatory signal through LFA-1. Crosslinking of anti-CR3 together with T-cell derived cytokines or LPS results in enhanced tissue factor response of mouse macrophages, indicating a signalling function for CR3.[139] In addition, crosslinked anti-LFA-1 or anti-CR3 antibodies directed against the α or β chain were able to induce cell-associated IL-2 expression on monocytes.[140]

Likewise, anti-β1 antibodies appeared to have costimulatory activity, suggesting that VLA molecules are also capable of "outside-in" signalling.[141] Similarly, immobilized purified fibronectin, or fragments thereof, were able to promote T-lymphocyte proliferation of CD4+ cells induced by anti-CD3 antibodies.[142-145] This costimulatory activity of fibronectin is primarily mediated through VLA-4 and VLA-5. The costimulatory activity of immobilized fibronectin was not affected by the soluble fibronectin present in the medium (serum), indicating that immobilized fibronectin was bound more avidly than soluble fibronectin. Other extracellular matrix components, like collagen and laminin, have also been shown to provide costimulatory signals together with CD3.[98,146] Roberts et al. have demonstrated that for production of IL-4 by a subset of γ/δ mouse T cells, both the vitronectin receptor αv/β3 and the TCR must be engaged by their respective ligands.[147]

Certain ligands of integrins are transmembrane proteins (VCAM-1, ICAM-1, ICAM-2) and are therefore potential candidates for "outside-in" signalling. It has been observed that these counter-receptors differ from integrins in that they do not express an active and inactive state of the receptor as described above,[80,81] implying that they do not serve as an "inside-out" receptor. However, despite this observation, it cannot be completely excluded that these molecules have an "outside-in" signalling function, although direct evidence is lacking. "Outside-in" signalling through adhesion receptors is not absolute necessarily regulated by intracellular messengers but may also be mediated by cytoskeletal components. Many of the receptors and presumably also their transmembrane ligands are linked to the cytoskeleton. Engagement of a receptor with its counterpart on an adjacent cell may alter the cell surface distribution of the receptor and thereby affect the cytoskeletal organization. Crosslinking of receptors, either by antibody or by immobilized ligand (soluble ligands generally do not have costimulatory activity), is apparently an absolute requirement in this process and may trigger the cell to respond (migration, cytokine production, proliferation, cytotoxic activity, etc).

ORGANIZATION OF ADHESION RECEPTORS AT THE CELL SURFACE AND THE ROLE OF THE CYTOSKELETON

As mentioned previously, integrins are transmembrane proteins and when these receptors bind to their ligands, for instance fibronectin or other extracellular matrix proteins, they become organized in so-called "focal adhesions" or "focal contacts."[148] In these focal contacts integrin receptors are clustered and linked to cytoskeletal elements (see Figure 7-3). Multiple receptor–ligand interactions in close proximity are thought to overcome the relatively low affinity of the receptor to its ligand and may result in stable adhesion.[95,149] The dynamics of the components that participate in focal adhesion have been studied by fluorescence recovery after photobleaching.[150,151] These experiments revealed retarded lateral diffusion of membrane proteins but also of cytoskeletal components, including talin, vinculin, alpha actinin, and actin, and show only a slow exchange between the proteins in the focal adhesions and the surrounding pool of cytoplasmic proteins. The integrins in the focal adhesions are important for anchoring of stress fibers to the plasma membrane but also regulate assembly and disasssembly of the attached actin filaments.[152,153] Talin can link integrin molecules to the cytoskeleton.[154] This interaction can be inhibited by a synthetic peptide that corresponds to a conserved sequence in the integrin beta chain[155] (Topley et al., unpublished data). Although this interaction is specific, it is of low affinity, suggesting the involvement of other proteins, such as vinculin, in the transduction of signals through integrin receptors into the interior of the cell.

Further evidence for the involvement of talin in linkage of integrins to cytoskeleton comes from observations of Kupfer et al.,[156,157] who showed that talin is concentrated at the site of adhesion between cytotoxic T lymphocytes and target cells and between antigen presenting B lymphocytes and helper T lymphocytes. Also, in platelets an activation-dependent change in the subcellular localization has been observed.[158] Similarly, Müller et al.[159] showed that binding of normal as well as Rous sarcoma virus transformed fibroblasts to substrate-bound fibronectin causes integrin–talin co-aggregation at the sites of membrane–extracellular matrix contact, which may thereby initiate the cytoskeletal events necessary for cell adhesion and spreading.

Various investigators have shown that phorbol esters, which activate protein kinase C, have a dramatic effect on cell adhesion and cytoskeleton organization. Generally, addition of PMA to epithelial cell cultures leads to a disruption of the cytoskeleton,[160–162] which is preceded by a loss of vinculin from focal adhesions.[163] An intriguing discrepancy exists between the effects of PMA on epithelial cells compared to hematopoietic cells. Whereas PMA results in disruption of focal adhesion in epithelial cells, it enhances cellular adhesion and association of integrin receptors with talin in lymphocytes.[164]

The presence of kinases in the adhesion of cells transformed by particular viruses[148] indicates that similar kinases may be involved in transmembrane signalling of normal cells. Elevated levels of phosphotyrosine have indeed been detected in the focal adhesions of normal cells.[165] Stimulation of granulocytes with the chemotactic peptide FMLP results in a rapid association of the chemotactic peptide–receptor complex with the cytoskeleton. This association is likely to be regulated by the activation and dissociation of the FMLP associated G protein.[166] Futhermore, alteration in $[Ca^{2+}]_i$ levels are associated with adhesion and migration of neutrophils.[167,168] Integrin-mediated adhesion generates multiple $[Ca^{2+}]_i$ transients, which may control events associated with cell movement and activation or priming of neutrophils to other stimuli ("outside-in" signalling).

REDUNDANCY

One important question that arises from the various adhesion pathways that have been discovered in the past decade is: Why are there so many different routes that can be employed for an apparently similar type of cellular interaction? There is no simple answer to this question. Indeed, it is sometimes not clear why for instance a CTL–target cell interaction is mediated by both the LFA-1/ICAM-1 pathway, the CD2/LFA-3 pathway,[111] and possibly also by VLA-4. However, on closer examination, it appears that LFA-1 may be of special importance during initiation of an immune response, whereas a fully activated cytotoxic effector cell is capable to lyse its target in the complete absence of LFA-1 (P. van de Wiel-van Kemenade, submitted). In addition, nonspecific cytotoxicity seems to depend more on LFA-1 than on CD2, giving further distinction to the differences between both adhesion pathways. A role for VLA-4 in cytotoxicity may be limited to certain specific T cell subsets. Similarly, VLA-4/VCAM-1 but also ELAM-1 interactions play a prominent role in lymphocyte binding to cytokine-activated endothelium, whereas LFA-1 is particularly involved in binding to resting endothelium and contributes less to adhesion to stimulated endothelium, despite the enhanced expression of ICAM-1 on cytokine activated endothelium (C. Vennegoor, submitted, unpublished results). These observations suggest that although it may seem that all these different adhesion molecules are participating in the same intercellular interactions, there are subtle differences, not only with respect to the type of interaction (migration, proliferation, antigen presentation, cytotoxicity), but also with respect to the activation state of the cell or a specific subset of cells. For instance, the role of ELAM-1 in lymphocyte adhesion to endothelium seems to be limited to a subpopulation of "memory" T lymphocytes of the skin.[55] Recent observations indicate that adhesion pathways may be linked to each other and act sequentially instead of in parallel (see below). In conclusion, although at first glance operation of several adhesion pathways in a specific cellular interaction may point to redundancy, subtle characteristics in each individual adhesion pathway can be distinguished on closer examination.

ADHESION CASCADES

As mentioned previously, several experiments indicate that adhesion molecules may operate successively. One example is the T cell receptor–antigen/ MHC interaction (see Figure 7-1), which can be regarded as an adhesion pathway (it mediates cell–cell contact). This interaction precedes the LFA-1 or VLA-4–6 interaction.[80,81,99] Similarly, a CD2/LFA-3 interaction may activate the LFA-1/ICAM-1 interaction,[80] although both pathways may also interact independently (P. van de Wiel-van Kemenade, submitted). Also, CD44, which may play a role in lymphocyte homing, can stimulate the LFA-1 adhesion pathway.[169] Other observations report the existence of adhesion cascades in myeloid cells. There is preliminary evidence that particularly the "rolling receptors" (selectins) act very early in the adhesion cascades. They enable rolling of neutrophils along the endothelial cells, thus reducing the speed of the cells in the circulation. They bring the neutrophils in close proximity of the endothelial cells, thus facilitating engagement of adhesion molecules (integrins) in adhesion and migration. ELAM-1-, but also GMP-140- mediated interactions can both promote CR3-mediated binding to endothelium or to a purified ICAM-1 substrate.[170–173] Furthermore, CR3-mediated adhesion is often associated with rapid downregulation of LECAM-1, presumably by proteolytic cleavage.[51] It is unknown if ELAM-1 interactions also induce this activity. It will be of great interest to resolve the underlying intracellular signals of these adhesion cascades, which may help us to understand the obviously complicated regulatory circuits that control cell adhesion and locomotion.

SYNOPSIS AND DISCUSSION

It becomes clear from the observations made to date that enormous insight into the structure of adhesion molecules has been gained, although important three-dimensional information is still largely lacking. In addition, receptors and ligands have been defined and it is known in which processes they operate. It is known that adhesion can be regulated by at least four different levels: 1) modulation of the number of receptor or ligand molecules expressed at the cell surface, 2) maturation/activation processes may modify adhesion molecules (glycosylation, expression of new epitopes) or enzymatically clip molecules from the cell surface, 3) specific activation of adhesion receptors can affect their ligand binding affinity; and 4) changes in the organization of the cytoskeleton and of adhesion receptors at the cell surface can affect cell adhesion.

Intracellular signalling pathways have hardly been investigated and form an interesting area of future research. In addition, a more thorough understanding of the role of the cations in the activation of integrin molecules may in part provide new answers on the regulation of these molecules. Another interesting but difficult area of research is the identification of adhesion cascades (for instance, how is ELAM-1 connected to engagement of CR3?). These questions show that despite all the structural information, understanding of the mechanisms that regulate adhesion and de-adhesion is still in its infancy. The most important answers concerning regulation of these intriguing molecules are still to come.

Acknowledgments: The authors thank Marie Anne van Halem for her help in preparing this manuscript. The work is in part supported by grants from the "Nierstichting" (Grant no. C87 724) and The Dutch Cancer Foundation (Grant no. NKI 87-5).

REFERENCES

1. Stoolman, L.M. (1989) *Cell* 56, 907–910.
2. Harlan, J.M. (1985) *Blood* 65, 513–525.
3. Springer, T.A. (1990) *Nature* 346, 425–434.
4. Larson, R.S., Corbi, A.L., Berman, L., and Springer, T.A. (1989) *J. Cell Biol.* 108, 703–712.
5. Dahms, N.M. and Hart, G.W. (1985) *J. Immunol.* 134, 3978–3986.
6. Van de Water, L., Aronson, D., and Braman, V. (1988) *Cancer Res.* 48, 5730–5737.
7. Symington, B.E., Symington, F.W., and Rohrschneider, L.R. (1989) *J. Biol. Chem.* 264, 13258–13266.
8. Jalkanen, S., Bargatze, R., Hsu, M., Wu, D. N., and Butcher, E.C. (1985) *Fed. Proc.* 44, 1177.
9. Simmons, D., Makgoba, M.W., and Seed, B. (1988). *Nature* 331, 624–627.
10. Dustin, M.L., Rothlein, R., Bhan, A.K., Dinarello, C.A., and Springer, T.A. (1986) *J. Immunol.* 137, 245–254.
11. Pober, J.S., Bevilacqua, M.P., Mendrick, D.L., Lapierre, L.A., Fiers, W., and Gimbrone, Jr., M.A. (1986) *J. Immunol.* 136, 1680–1687.
12. Pober, J.S., Gimbrone, Jr. M.A., Lapierre, L.A., Mendrick, D.L., Fiers, W., Rothlein, R., and Springer, T.A. (1986) *J. Immunol.* 137, 1893–1896.
13. Pober, J.S., Lapierre, L.A., Stolpen, A.H., Brock, T.A., Springer, T.A., Fiers, W., Bevilacqua, M.P., Mendrick, D.L., and Gimbrone, Jr., M.A. (1987) *J. Immunol.* 138, 3319–3324.
14. Renkonen, R. (1989) *Scand. J. Immunol.* 29, 717–721.
15. Cotran, R.C., Gimbrone, Jr., M. A., Bevilacqua, M.P., Mendrick, D.L., and Pober, J.S. (1986) *J. Exp. Med.* 164, 661–666.
16. Makgoba, M.W., Sanders, M.E., Luce, G.E., Dustin, M., Springer, T.A., Clark, E.A., Mannoni, P., and Shaw, S. (1988) *Nature* 331, 86–88.
17. Dustin, M.L., and Springer, T.A. (1988) *J. Cell Biol.* 107, 321–333.
18. Staunton, D.E., Dustin, M.L., and Springer, T.A. (1989) *Nature* 339, 61–64.
19. Shimizu, Y., Van Seventer, G.A., Horgan, K.J., and Shaw, S. (1990) *Nature* 345, 250–253.
20. Arnaout, M.A., Spits, H., Terhorst, C., Pitt, J., and Todd III, R.F. (1984) *J. Clin. Invest.* 74, 1291–1300.
21. Arnaout, M.A., Wang, E.A., Clark, S.C., and Sieff, C.A. (1986). *J. Clin. Invest.* 78, 597–601.
22. Gamble, J.R., Harlan, J.M., Klebanoff, S.J., and Vadas, J.A. (1985) *Proc. Natl. Acad. Sci. USA* 82, 8667–8671.
23. Hibbs, M.S. and Bainton, D.F. (1989) *J. Clin. Invest.* 84, 1395–1402.
24. Buyon, J.P., Abramson, S.B., Philips, M.P., Slade, S.G., Ross, G.D., Weissman, G., and Winchester, R.J. (1988) *J. Immunol.* 140, 3156–3160.
25. Vedder, N.B. and Harlan, J.M. (1988) *J. Clin. Invest.* 81, 939–944.
26. Hemler, M.E., Elices, M.J., Parker, C., and Takad, Y. (1990) *Immunol. Rev.* 114, 45–65.
27. Hynes, R.O. (1987) *Cell* 48, 549–554.
28. Osborn, L., Hession, C., Tizard, T., Vassallo, C., Luhowskyj, S., Chi-Rosso, G., and Lobb, R. (1989) *Cell* 59, 1203–1211.

29. Price, G.E. and Bevilacqua, M.P. (1989) *Science* 246, 1303–1306.
30. Holzmann, B., McIntyre, B.W., and Weissman, I.L. (1989) *Cell* 56, 37–46.
30a. Holzmann, B., and Weissman, I.L. (1989) *EMBO Journal* 8, 1735–1741.
31. Hemler, M.E., Huang, C., Takada, Y., Schwarz, L., Strominger, J.L., and Clabby, M.L. (1987) *J. Biol. Chem.* 262, 11478–11485.
32. Hemler, M.E., Takada, Y., Elices, M., and Crouse, C. (1989) in *Leukocyte Adhesion Molecules* (Springer, T.A., Anderson, D.A., Rosenthal, A.S., and Rothlein, R., eds.) pp. 44–57, Springer-Verlag, New York.
33. Wayner, E.A., Garcia-Pardo, A., Humphries, M.J., McDonald, J.A., and Carter, W.G. (1989) *J. Cell Biol.* 109, 1321–1330.
34. Sanders, M.E., Makgoba, M.W., Sharrow, S.O., Stephany, D., Springer, T.A., Young, H.A., and Shaw, S. (1988) *J. Immunol.* 140, 1401–1407.
35. Spits, H., Van Schooten, W., Keizer, H., Van Seventer, G., Van de Rijn, M., Terhorst, C., and De Vries, J.E. (1986) *Science* 232, 403–405.
36. Hughes, C.C.W., Savage, C.O.S., and Pober, J.S. (1990) *J. Exp. Med.* 171, 1453–1467.
37. Buckle, A.M. and Hogg, N. (1991) *Eur. J. Immunol.* In press.
38. Heino, J., Ignotz, R.A., Hemler, M.E., Crouse, C., and Massagué, J. (1989) *J. Biol. Chem.* 264, 380–388.
39. Ignotz, R.A., Heino, J., and Massagué J. (1989) *J. Biol. Chem.* 264, 389–392.
40. Gallatin, M.W., Weissman, I.L., and Butcher, E.C. (1983) *Nature* 304, 30–34.
41. Lasky, L.A., Singer, M.S., Yednock, T.A., Dowbenko, D., Fennie, C., Rodriguez, H., Nguyen, T., Stachel, S., and Rosen, S.D. (1989) *Cell* 56, 1045–1055.
42. Siegelman, M.H. and Weissman, I.L. (1989) *Proc. Natl. Acad. Sci. USA* 86, 5562–5566.
43. Bowen, B.R., Nguyen, T., and Lasky, L.A. (1989) *J. Cell Biol.* 109, 421–438.
44. Tedder, T.F., Isaacs, C.M., Ernst, T.J., Demetri, G.D., Adler, D.A., and Disteche, C.M. (1989) *J. Exp. Med.* 170, 123–133.
45. Bevilacqua, M.P., Stengelin, S., Gimbrone, Jr., M.A., and Seed, B. (1989) *Science* 243, 1160–1165.
46. Johnston, G.I., Cook, R.G., and McEver, R.P. (1989) *Cell* 563, 1033–1044.
47. Larsen, E., Celi, A., Gilbert, G.E., Furie, B.C., Eban, J.K., Bonfati, R., Wagner, D.D., and Furie, B. (1989) *Cell* 59, 305–312.
48. Yednock, T.A. and Rosen, S.D. (1989) *Adv. Immunol.* 44, 313–378.
49. Phillips, M.L., Nudelman, E., Gaeta, F.C.A., Perez, M., Singhal, A.K., Hakomori, S.I., and Paulson, J.C. (1990) *Science* 250, 1130–1132.
50. Smith, C.W., Kishimoto, T.K., Abbass, O., Hughes, B., Rothlein, R., McIntire, L.V., Butcher, E., and Anderson, D.C. (1991) *J. Clin. Invest.* 87, 609–618.
51. Kishimoto, T.K., Jutila, M.A., Berg, E.L., and Butcher, E.C. (1989) *Science* 245, 1238–1241.
52. Tedder T.E., Matsuyama, T., Rothstein, D., Schlossman, S.F., and Morimoto, C. (1990) *Eur. J. Immunol.* 20, 1351–1355.
53. Kishimoto, T.K., Anderson, D.C., Butcher, E., and Smith, C.W. (1990) *J. Leukocyte Biol.* 48 (Suppl. 1), 96–99.
54. Shimizu, Y., Shaw, S., Graber, N., Gopal, T.V., Horgan, K.J., van Seventer, G.A., and Newman, W. (1991) *Nature* 349, 799–802.
55. Picker, L.J., Kishimoto, T.K., Smith, C.W., Warnock, R.A., and Butcher, E.C. (1991) *Nature* 349, 796–799.

56. Bevilacqua, M.P., Pober, J.S., Mendrick, D.L., Cotran, R.S., and Gimbrone, M.A. (1987) *Proc. Natl. Acad. Sci. USA* 84, 9238–9242.
57. Luscinskas, F.W., Brock, A.F., Arnaout, M.A., and Gimbrone, M.A. (1989) *J. Immunol.* 143, 3318–3324.
58. Bonfanti, R., Furie, B.C., Furie, B., and Wagner, D.D. (1989) *Blood* 73, 1109–1112.
59. Larsen, E., Celi, A., Gilbert, G.E., Furie, B.C., Erban, J.K., Bonfanti, R., Wagner, D.D., and Furie, B. (1989) *Cell* 59, 305–312.
60. Geng, J.G., Bevilacqua, M.P., Moore, K.L., McIntyre, T.M., Prescott, S.M., Kim, J.M., Bliss, G.A., Zimmerman, G.A., and McEver, R.P. (1990) *Nature* 343, 757–760.
61. Larsen, E., Plabrica, T., Sajer S., Gilbert, G.E., Wagner, D.D., Furie, B.C., and Furie, B. (1990) *Cell* 63, 467–474.
62. Gamble, J.R., Skinner, M.P., Berndt, M.C., and Vadas, M.A. (1990) *Science* 249, 414–417.
63. Tangelder, G.J. and Arfors, K.E. (1991) *Blood* 77, 1565–1571.
64. Patarroyo, M., Yogeeswaran, G., Biberfeld, P., Klein, E., and Klein, G. (1982) *Int. J. Cancer* 30, 707–717.
65. Patarroyo, M., Prieto, J., Ernberg, I., and Gahmberg, C.G. (1988) *Int. J. Cancer* 41, 901–907.
66. Clayberger, C., Wright, A., Medeiros, L.J., Koller, T.D., Link, M.P., Smith, S.D., Warnke, R.A., and Krensky, A.M. (1987) *Lancet* ii, 533–536.
67. Billaud, M., Rousset, F., Calender, A., Cordier, M., Aubry, J.P., Laisse, V., and Lenoir, G.M. (1990) *Blood* 75, 1827–1833.
68. Wang, D., Liebowitz, D., Wang, F., Gregory, C., Rickinson, A., Larson, R.S., Springer, T.A., and Kieff, E. (1988). *J. Virol.* 62, 4173–4184.
69. Strassman, G., Springer, T.A., Haskill, S.J., Miraglia, C.C., Lanier, L.L., and Adams, D. (1985) *Cell. Immunol.* 94, 265–275.
70. Bender, J.R., Tachett, L., and Pardi, R. (1991) *Transpl. Proc.* 23, 99–101.
71. Martz, E. (1980) *J. Cell Biol.* 84, 584–598.
72. Arnaout, M.A. (1990) *Immunol. Rev.* 114, 144–180.
73. Patarroyo, M. and Jondal, M. (1985) *Immunobiology* 170, 305–319.
74. Patarroyo, M., Beatty, P.G., Fabre, J.W., and Gahmberg, C.G. (1985) *Scand. J. Immunol.* 22, 171–182.
75. Wright, S.D. and Meyer, B.C. (1986) *J. Immunol.* 136, 1759–1764.
76. Harlan, J.M., Schwartz, B.R., Reidy, M.A., Schwartz, S.M., Ochs, H.D., and Harker, L.A. (1985) *Lab. Invest.* 52, 141–150.
77. Marguerie, G.A., Plow, E.F., and Edgington, T.S. (1979) *J. Biol. Chem.* 254, 5357–5363.
78. Bennett, J.S. and Vilaire, G. (1979) *J. Clin. Invest.* 64, 1393–1401.
79. Shattil, S.J. and Brass, L.F. (1987) *J. Biol. Chem.* 262, 992–1000.
80. Van Kooyk, Y., Van de Wiel-van Kemenade, P., Weder, P., Kuijpers, T.W., and Figdor C.G. (1989) *Nature* 342, 811–813.
81. Dustin, M.L. and Springer, T.A. (1989) *Nature* 341, 619–624.
82. Tamatani, T., Kotani, M., Tanaka, T., and Miyasaka, M. (1991) *Eur. J. Immunol.* 21, 855–858.
83. Axelsson, B., Youseffi-Etemad, R., Hammarström, S., and Perlmann, P. (1988) *J. Immunol.* 141, 2912–2917.

84. Koopman, G., van Kooyk, Y., de Graaff, M., Meyer, C.J.L.M., Figdor, C.G., and Pals, S.T. (1990) *J. Immunol.* 145, 3589–3593.
85. Barrett, T.B., Shu, G., and Clark, E.A. (1991) *J. Immunol.* 146, 1722–1729.
86. Lauener, R.P., Geha, R.S., and Vercelli, D. (1990) *J. Immunol.* 145, 1390–1394.
87. Forsyth, K.D., Simpson, A.C., and Levinsky, R.J. (1989) *Eur. J. Immunol.* 19, 1331–1334.
88. Mourad, W., Geha, R.S., and Chatila, T. (1990) *J. Exp. Med.* 172, 1513–1516.
89. Peters, P.J., Geuze, H.J., Van der Donk, H.A., Slot, J.W., Griffith, J.M., Stam, N.J., Clevers, H.C., and Borst, J. (1989) *Eur. J. Immunol.* 19, 1469–1475.
90. Sarkadi, B., Tordai, A., Müleer, M., and Gardos, G. (1990) *Mol. Immunol.* 27, 1297–1306.
91. Marano, N., Holowka, D., and Baird, B. (1989) *J. Immunol.* 143, 931–938.
92. Alexander, D.R. and Cantrell, D.A. (1989) *Immunol. Today 10*, 200–205.
93. Figdor, C.G., Van Kooyk, Y., and Keizer, G.D. (1990) *Immunol. Today* 11, 277–280.
94. Altieri, D.C., Wiltse, W.L., and Edgington, T.S. (1990) *J. Immunol.* 145, 662–670.
95. Detmers, P.A., Wright, S.D., Olsen, E., Kimball, B., and Cohan, Z.A. (1987) *J. Cell Biol.* 105, 1137–1135.
96. Phillips, M.R., Buyon, J.P., Winchester, R., Weissmann, G., and Abramson, S.B. (1988) *J. Clin. Invest.* 82, 495–501.
97. Detmers, P.A., Lo, S.K., Olsen-Egbert, E., Walz, A., Baggiolini, M., and Cohn, Z.A. (1990) *J. Exp. Med.* 171, 1155–1162.
98. Wilkins, J.A., Stupack, D., Stewart, S., and Caixia, S. (1991) *Eur. J. Immunol.* 21, 517–522.
99. Shimizu, Y., van Seventer, G.A., Horgan, K.J., and Shaw, S. (1990) *Immunol. Rev.* 114, 109–143.
100. Pytela, R.P., Pierschbacher, M.D., Ginsberg, M.H., Plow, E.F., and Ruoslahti, E. (1986) *Science* 231, 1559–1562.
101. Lam, S.C.T., Plow, E.F., Smith, M.A., Andrieux, A., Ryckwaert, J.J., Marguerie, G., and Ginsberg, M.H. (1987) *J. Biol. Chem.* 262, 947–950.
102. Parise, L.V., Helgerson, S.L., Steiner, B., Nannizzi, L., and Phillips, D.R. (1987) *J. Biol. Chem.* 262, 12597–12602.
103. Frelinger III, A., Cohen, I., Plow, E.F., Smith, M.A., Roberts, J., Lam, S.C.T., and Ginsberg, M.H. (1988) *J. Biol. Chem.* 263, 12397–12402.
104. O'Toole, T.E., Loftus, J.C., Du, X., Glass, A. A., Ruggeri, Z.M., Shattil, S.J., Flow, E.F., and Ginsberg, M.H. (1990) *Cell Regulation* 1, 883–893.
105. Chatila, T.A. and Geha, R.S. (1988) *J. Immunol.* 144, 191–197.
106. Buyon, J.P., Slade, S.G., Reibman, J., Abramson, S.B., Philips, M.R., Weissman, G., and Winchester, R. (1990) *J. Immunol.* 144, 191–197.
107. Chatila, T.A., Geha, S., and Arnaout, M.A. (1989) *J. Cell Biol.* 109, 3435–3444.
108. Hibbs, M.L., Xu, H., Stacker, S.A., and Springer, T.A. (1991) *Science* 251, 1611–1613.
109. Merrill, J.T., Slade, S.G., Weissmann, G., Winchester, R., and Buyon, J.P. (1990) *J. Immunol.* 145, 2608–2615.

110. Imboden, J.B. and Stobo, J.D. (1985) *J. Exp. Med.* 161, 446–456.
111. Krensky, A.M., Sanchez-Madrid, F., Robbins, E., Nagy, J., Springer, T.A., and Burakoff, S.J. (1983) *J. Immunol.* 131, 611–616.
112. Klausner, R.D. and Samelson, L.E. (1991) *Cell* 64, 875–878.
113. Shimizu, Y., Van Seventer, G.A., Siraganian, R., Wahl, L., and Shaw, S. (1989) *J. Immunol.* 143, 2457–2463.
114. Rothlein, R. and Springer, T.A. (1986) *J. Exp. Med.* 163, 1132–1142.
115. Keizer, G.D., Visser, W., Vliem, M., and Figdor, C.G. (1988) *J. Immunol.* 140, 1393–1400.
116. Van Kooyk, Y., Weder, P., Hogervorst, F., Verhoeven, A.J., Van Seventer, G., Te Velde, A.A., Borst, J., Keizer, G.D., and Figdor, C.G. (1991) *J. Cell Biol.* 112, 345–354.
117. Haskard, D., Cavender, D., Beatty, P., Springer, T.A., and Ziff, M. (1986) *J. Immunol.* 137, 2901–2906.
118. Dransfield, I. and Hogg, N. (1989) *EMBO J.* 12, 3759–3765.
119. Dransfield, I., Buckle, A.M., and Hogg, N. (1990) *Immunol. Rev.* 114, 29–44.
120. Larson, R.S. and Springer, T.A. (1990) *Immunol. Rev.* 114, 181–270.
121. Takada, Y. and Hemler, M.E. (1989) *J. Cell Biol.* 109, 397–407.
122. Staatz, W.D., Raqjpara, S.M., Wayner, E.A., Carter, W.G., and Santoro, S.A. (1989) *J. Cell Biol.* 108, 1917–1924.
123. Smith, J.W. and Cheresh, D.A. (1990) *J. Biol. Chem.* 265, 2168–2172
124. D'Souza, S.E., Ginsberg, M.H., Burke, T.A., and Plow, E.F. (1990) *J. Biol. Chem.* 265, 3440–3446.
125. D'Souza, S.E., Ginsberg, M.H., Matsueda, G.R., and Plow, E.F. (1991) *Nature* 350, 66–68.
126. Frelinger III, A.L., Cohen, I., Plow, E.F., Smith, M.A., Roberts, J., Lam, S.C.T., and Ginsberg, M.H. (1990) *J. Biol. Chem.* 265, 6346–6352.
127. Kouns, W.C., Fox, C.F., Lamareaux, W.J., Coons, L.B., and Jennings, L.K. (1989) *Blood* 74 (suppl. 1), 91a.
128. Larson, R.S., Hibbs, M.L., and Springer, T.A. (1990) *Cell Regulation* 1, 359–367.
129. Gulino, D., Ryckewaert, J.J., Andrieux, A., Rabiet, M.J., and Marguerie, G. (1990) *J. Biol. Chem.* 265, 9575–9581.
130. Bednarczyk, H.L. and McIntyre, B.W. (1990) *J. Immunol.* 144, 777–784.
131. Campanero, M.R., Pulido, R., Ursa, M.A., Rodriguez-Moya, M., de Landázuri, M.O., and Sánchez-Madrid, F. (1990) *J. Cell Biol.* 110, 2157–2165.
132. Clayberger, C., Krensky, A.M., McIntyre, B.W., Koller, T.D., Parham, P., Brodsky, F., Linn, D.J., and Evans, E.L. (1987) *J. Immunol.* 138, 1510–1514.
133. Neugebauer, K.M. and Reichardt, L.F. (1991) *Nature* 350, 68–71.
134. Pardi, R., Bender, J.R., Dettori, C., Gianazza, E., and Engleman, E. G. (1989) *J. Immunol.* 143, 3157–3166.
135. Wacholtz, M.C., Patel, S.S., and Lipsky, P.E. (1989) *J. Exp. Med.* 170, 431–448.
136. Pircher, H., Groscurth, P., Baumhutter, S., Aguet, M., Zinkernagel, R.M., and Hengartner, H. (1986) *Eur. J. Immunol.* 16, 172–181.
137. Richter, J., Ng-Sikorski, J., Olsson, I., and Andersson, T. (1990) *Proc. Natl. Acad. Sci. USA* 87, 9472–9476.
138. Van Noesel, C., Miedema, F., Brouwer, M., De Rie, M.A., Aarden, L.A., and Van Lier, R.A.W. (1988) *Nature* 333, 850–952.

139. Fan, S.T. and Edgington, T.S. (1991) *J. Clin. Invest.* 87, 50–57.
140. Couturier, C., Haeffner-Cavaillon, N., Weiss, L., Fischer, E., and Kazatchkine, M.D. (1990) *Eur. J. Immunol.* 20, 999–1005.
141. Yamada, A., Nikaido, T., Nojima, Y., Schlossman, S.F., and Morimoto, C. (1991) *Eur. J. Immunol.* 21, 319–325.
142. Davis, L.S., Oppenheimer-Marks, N., Bednarczyk, J.L., McIntyre, B.W., and Lipsky, P.E. (1990) *J. Immunol.* 145, 785–793.
143. Matsuyama, T., Yamada, A., Kay, J., Yamada, K.M., Akiyama, S.K., Schlossman, S.F., and Morimoto, C. (1989) *J. Exp. Med.* 170, 1133–1148.
144. Nojima, Y., Humphries, M.J., Mould, A.P., Komoriya, A., Yamada, K.M., Schlossman, S.F., and Morimoto, C. (1990) *J. Exp. Med.* 172, 1185–1192.
145. Yamada, A., Nikaido, T., Nojima, Y., Schlossman, S.F., and Morimoto, C. (1991) *J. Immunol.* 146, 53–56.
146. Dang, N.H., Torimoto, Y., Schlossman, S.F., Morimoto, C. (1990) *J. Exp. Med.* 172, 649–652.
147. Roberts, K., Yokoyama, W.M., Kehn, P.J., and Shevach, E.M. (1991) *J. Exp. Med.* 173, 231–240.
148. Burridge, K., Fath, K., Kelly, T., Nuckolls, G., and Turner, C. (1988) *Ann. Rev. Cell Biol.* 4, 487–525
149. Hermanowski-Vosatka, A., Detmers, P.A., Gotze, O., Silverstein, S.C., and Wright, S.D. (1988) *J. Biol. Chem.* 263, 17822–17827.
150. Geiger, B. (1982) *J. Mol. Biol.* 159, 685–701.
151. Geiger, B., Avnur, Z., Kreis, T.E., and Schlessinger, J. (1984) *Cell Muscle Motil.* 5, 195–234.
152. Kreis, T.E., Geiger, B., and Schlessinger, J. (1982) *Cell* 29, 835–845.
153. Wang, E. (1984) *J. Cell Biol.* 99, 1478–1485.
154. Horwitz, A., Duggan, K., Buck, C., Beckerle, M.C., and Burridge, K. (1986) *Nature* 320, 531–533.
155. Buck, C.A. and Horwitz, A.F. (1987) *Ann. Rev. Cell Biol.* 3, 179–205.
156. Kupfer, A., Singer, S.J., and Dennert, G. (1986) *J. Exp. Med.* 163, 489–498.
157. Kupfer, A., Swain, S.L., and Singer, S.J. (1987) *J. Exp. Med.* 165, 1565–1580.
158. Beckerle, M.C., Miller, D.E., Bertagnolli, M.E., and Locke, S.J. (1989) *J. Cell Biol.* 109, 3333–3346.
159. Müller, S.C., Kelly, T., Dai, M., Dai, H., and Chen, W.T. (1989) *J. Cell Biol.* 109, 3455–3464.
160. Werth, D.K., Niedel, J.E., and Pastan, I. (1983) *J. Biol. Chem.* 258, 11423–11426.
161. Litchfield, D.W. and Ball, E.M. (1986) *Biochem. Biophys. Res. Comm.* 134, 1276–1283.
162. Beckerle, M.C., O'Halloran, T., Earp, S., and Burridge, K. (1985) *J. Cell Biol.* 101, 411a.
163. Werth, D.K. and Pastan, I. (1984) *J. Biol. Chem.* 259, 5264–5270.
164. Burn, P., Kupfer, A., and Singer, S.J. (1988) *Proc. Natl. Acad. Sci. USA* 85, 497–501.
165. Maher, P.A. Pasquale, E.B., Wang, J.Y.J., and Singer, S.J. (1985) *Proc. Natl. Acad. Sci. USA* 82, 6575–6580.
166. Särndahl, E., Lindroth, M., Bengtsson, T., Fällman, M., Gustavsson, J., Stendahl, O., and Andersson, T. (1989) *J. Cell Biol.* 109, 2791–2799.

167. Jaconi, J.E.E., Theler, J.M., Schlegel, W., Appel, R.D., Wright, S.D., and Lew, P.D. (1991) *J. Cell Biol.* 112, 1249–1257.
168. Marks, P.W., Hendey, B., and Maxfield, F.R. (1991) *J. Cell Biol.* 112, 149–158.
169. Pals, S.T., Hogervorst, F., Keizer, G.D., Thepen, T., Horst, E., and Figdor, C.G. (1989) *J. Immunol.* 143, 851–857.
170. Lawrence, M.B. and Springer, T.A. (1991) *Cell* 65, 859–873.
171. Kuijpers, T.W., Hakkert, B.C., Hoogerwerf, M., Leeuwenberg, J.F.M., and Roos, D. (1991) *J. Clin. Invest.* 147, 1369–1376.
172. Lawrence, M.B., Smith, C.W., Eskin, S.G., and McIntire, L.V. (1990) *Blood* 75, 227–237.
173. Lo, S.K., Lee, S., Ramos, R.A., Lobb, R., Rosa, M., Chi-Rosso, G., and Wright, S.D. (1991) *J. Exp. Med.* 173, 1493–1500.
174. Strassman, G., Springer, T.A., and Adams D.O. (1985) *J. Immunol.* 135, 147–153.
175. Paul, W.E. (1991) *Blood* 77, 1859–1870.

CHAPTER 8

Outlook for the Future

David Y. Liu, John M. Harlan

Intercellular adhesion represents a discrete and significant event in the coordinated sequence of events leading to the pathology of inflammatory disease. This book has described in abundant detail the various *in vitro* and *in vivo* parameters that influence leukocyte binding to and traversing of endothelial cells. Amidst the startling explosion of published reports aimed at providing more information about the mechanisms of cell adhesion, this body of information was integrated in the most comprehensive and current manner. Indicative of how rapidly this field is moving, there have been several studies published in the last few months that merit a brief discussion.

The subject of activation of adhesion molecules and signals transduced through adhesion molecules as described in Chapter 7 has been the subject of most of the recently published articles. Activation of T lymphocytes[1,2] and fibroblasts[3] can proceed through the binding of integrins and homing receptors. The induction of CD18-mediated adhesion can occur through the binding of manganese ions[4] and endothelial leukocyte adhesion molecule-1 (ELAM-1).[5] Several reports described sialyl-Lewis A (SLe^a), an isomer of SLe^x in which the sialic acid and fucose linkages are reversed, as a ligand for

ELAM-1,[6-8] although SLe[a] may be more important for metastasis than for leukocyte recruitment.[6] Stimulation of ELAM-1 expression was reported to result from gene transcription through the activation of transcription factor (NF-κB) binding sites.[9] The functional domains of L-selectin[10] as well as the genomic structure of the vascular cell adhesion molecule-1 (VCAM-1) gene[11] were mapped. An anti-CD18 antibody was shown to be effective in blocking polymorphonucleocyte (PMN)-mediated acute lung injury in a pig model of gram-negative sepsis.[12] Endothelium-derived nitric oxide (NO) was suggested to be an endogenous inhibitor of leukocyte adhesion *in vivo*.[13] Finally, a new nomenclature for the selectin family of adhesion receptors has been recently adopted to facilitate the flow of information concerning these proteins and to reflect the functional significance of carbohydrate binding activity of these proteins[14] (see Glossary).

Early in the development of our understanding of leukocyte adhesion to endothelial cells, a question often asked concerned the apparent redundancy of functional adhesion receptors. It has become clear that specific groups of molecules have particular functions, and the results of many recent studies suggest that the process of leukocyte extravasation from the circulation can be divided into three arenas of adhesion molecule activity. This is depicted in Figures 5-3 and 8-1. In the initial phases of inflammation, there is a transient type of adhesion characterized by a rolling action of neutrophils along the venular wall. This adhesion event is mediated by members of the selectin family, as described in Chapters 2–6. The next stage of extravasation involves a stationary, firm type of adhesion in which activation of leukocytes plays a greater role, as does the integrin family of adhesion molecules. This is the immediate prelude to actual transmigration of leukocytes and is delineated in Chapters 1, 4, 5, and 6. Transendothelial migration proceeds by a CD18 integrin-dependent pathway with, perhaps, a role for ELAM-1

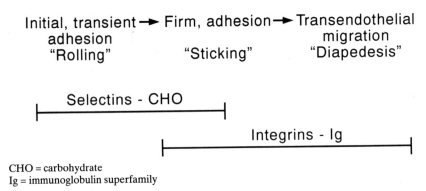

Figure 8-1. Leukocyte–endothelial cell interaction.

molecules as activators of the CD18 pathway.[5] Chapter 5 summarizes the results of studies on the role of adhesion molecules in transmigration.

The significance of activation of cells for increased adhesion should not be viewed simply in terms of increased expression of adhesion molecules but also in terms of increased avidity of adhesion molecules (Figure 8-2). Chapter 7 clearly presents evidence of the importance of the latter phenomenon in determining overall adhesivity of the system as do the more recent publications described above.[4,5] The integrin receptors, in particular, are most susceptible to changes in avidity as a result of cellular activation through the binding of adhesion molecules by their counter-receptors. Furthermore, the specificity or selectivity of adhesion can be influenced markedly by stimulus-specific activation, deactivation, or inhibition of cellular function as manifested in the expression of active adhesion molecules (Figure 8-3). This, of course, will depend heavily on the presence of appropriate endogenous mediators, particularly cytokines, as described in Chapter 4.

The basic premise for these important studies goes beyond the quest for scientific breakthroughs and extends into the realm of therapeutic intervention. It is obvious from the many studies described in Chapter 6 that blocking intercellular adhesion is an effective method for preventing the destructive phases of inflammation. Perhaps, the attention of the scientific and medical communities will no longer be focused on whether adhesion is an appropriate target for drug development but whether there are other molecular targets besides ligand-binding sites for interfering with adhesion, such as signal

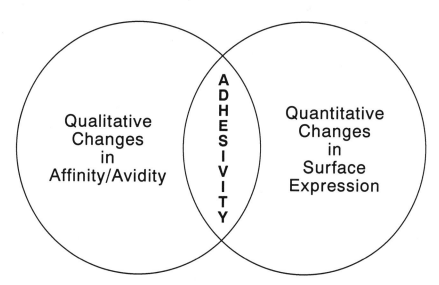

Figure 8-2. Adhesivity is determined by qualitative and quantitative changes.

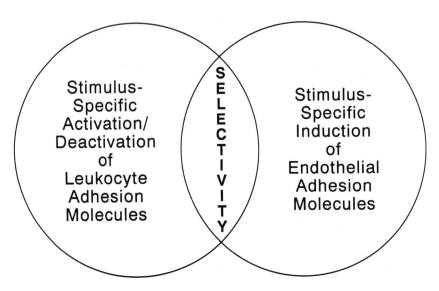

Figure 8-3. Determinants of selective recruitment of leukocyte subpopulations.

transduction pathways either leading to the expression of active adhesion molecules or resulting from the engagement of adhesion molecules. Finally, it remains to be seen whether there are more effective and practical ways of intervening other than through the use of monoclonal antibodies, such as soluble or chimeric adhesion receptors (and their counter-receptors), peptides based on binding site epitopes, small molecules (that is, carbohydrates based on selectin specificity and organic analogues based on any active structure), and antisense oligonucleotides to prevent adhesion gene expression. Although several important issues still need to be resolved, such as the nature of the selectin family counter-receptors, elucidation of adhesion-associated signal transduction pathways, and the specific role of adhesion molecules in particular inflammatory diseases, there is ample justification for pursuing the development of anti-inflammatory therapeutics that are based on the suppression of leukocyte adherence to and transmigration through the endothelium.

REFERENCES

1. Damle, N.K. and Aruffo, A. (1991) *Proc. Natl. Acad. Sci. USA* 88, 6403–6407.
2. Seth, A., Gote, L., Nagarkatti, M., and Nagarkatti, P.S. (1991) *Proc. Natl. Acad. Sci. USA* 88, 7877–7881.
3. Schwartz, M.A., Lechene, C., and Ingber, D. (1991) *Proc. Natl. Acad. Sci. USA* 88, 7849–7853.
4. Altieri, D. (1991) *J. Immunol.* 147, 1891–1898.

5. Kuijpers, T.W., Hakkert, B.C., Hoogerwerf, M., Leeuwenberg, J.F.M., and Roos, D. (1991) *J. Immunol.* 147, 1369–1376.
6. Berg, E.L., Robinson, M.K., Mansson, O., Butcher, E.C., and Magnani, J.L. (1991) *J. Biol. Chem.* 266, 14869–14872.
7. Takada, A., Ohmori, K., Takahashi, N., Kiyotaka, T., Yago, A., Zenita, K., Hasegawa, A., and Kannagi, R. (1991) *Biochem. Biophys. Res. Comm.* In Press.
8. Tyrell, D., James, P., Rao, N., Foxall, C., Abbas, S., Dasgupta, F., Nashed, M., Hasegawa, A., Kiso, M., Asa, D., Kidd, J., and Brandley, B. (1991) *Proc. Natl. Acad. Sci. USA.* In Press.
9. Montgomery, K.F., Osborn, L., Hession, C., Tizard, R., Goff, D., Vassallo, C., Tarr, P.I., Bomsztyk, K., Lobb, R., Harlan, J.M., and Pohlman, T.H. (1991) *Proc. Natl. Acad. Sci. USA* 88, 6523–6527.
10. Kansas, G.S., Spertini, O., Stoolman, L.M., and Tedder, T.F. (1991) *J. Cell Biol.* 114, 351–358.
11. Cybulsky, M.I., Fries, J.W.U., Williams, A.J., Sultan, P., Eddy, R., Byers, M., Shows, T., Gimbrone, Jr., M.A., and Collins, T. (1991) *Proc. Natl. Acad. Sci. USA* 88, 7859–7863.
12. Walsh, C.J., Carey, P.D., Cook, D.J., Bechard, D.E., Fowler, A.A., and Sugerman, H.J. (1991) *Surgery* 110, 205–211.
13. Kubes, P., Suzuki, M., and Granger, D. N. (1991) *Proc. Natl. Acad. Sci. USA* 88, 4651–4655.
14. Nomenclature Committee (1991) *Cell* 67, 233.

APPENDIX

In Vitro Adhesion Assay

David Y. Liu, Zehra Kaymakcalan

LEUKOCYTE ISOLATION

Human leukocytes are obtained from the venous blood of healthy adult volunteers. The procedure for polymorphonucleocyte (PMN) isolation is a modification of a previously described method.[1] In brief, blood is collected into heparin coated tubes and 20 ml is layered directly onto a Ficoll-Hypaque double density gradient (20 ml of 1119 and 10 ml of Histopaque 1077 [Sigma]) for centrifugation at $700 \times g$ for 30 minutes. PMN are collected from the 1077/1119 interphase, washed twice in RPMI (JRH Biosciences), and then the erythrocytes are removed by hypotonic lysis. The final cell pellet (95%–98% PMN) is resuspended in Hanks Buffered Salt Solution (HBSS) for radiolabeling.

Human monocytes can be isolated from peripheral blood by density gradient centrifugation[2] or by countercurrent centrifugal elutriation.[3] Human eosinophils can be purified on metrizamide[4] or percoll[5] gradients. When appropriate, human cell lines can be examined for their ability to adhere to endothelial cells (EC), such as the promyelocytic HL-60 line[6] and histiocytic U-937 line.[7]

ENDOTHELIAL CELLS

Human umbilical cords are obtained as a courtesy from hospitals (labor and delivery section). EC are isolated from human umbilical cords by mild collagenase (Worthington, Freehold, NJ) digestion as previously described.[8,9] Cells from individual cords are grown in flasks coated with 0.2% gelatin (Sigma) in the following growth medium: M199 with 25 mM Hepes (JRH, Lenexa, KS) plus 20% heat-inactivated FCS (JRH), 2 mM glutamine (Irvine Scientific, Santa Ana, CA), 60 µg/ml sodium heparin (140 U/mg)(Sigma),[10] and 50 µg/ml bovine hypothalamus (JRH, Lenexa, KS) extract as a source of endothelial cell growth factor.[11] Cultured umbilical vein endothelial cells exhibit a typical cobblestone morphology at confluence, stain positively for factor VIII antigen by indirect immunofluorescence, and are not used after the fifth passage. For subculture, confluent cells are rinsed once with 0.2% EDTA in PBS, harvested by treatment with 0.25% trypsin (GIBCO) for one minute at room temperature, and diluted 1:4 in fresh growth medium. Certain studies may warrant the use of only freshly harvested EC.[12] Furthermore, significant empirical differences between various types of EC, such as large vessel venous, large vessel arterial, and microvascular capillary EC, have been documented.[13]

LEUKOCYTE/ENDOTHELIAL CELL STATIC ADHESION ASSAY

The assay is a modification of a previously described procedure.[14] Monolayers of EC are grown in flat-bottomed 96-well plates (Corning) that had been precoated with 6.4 µg/ml human plasma fibronectin (New York Blood Center, NY) for 30 minutes at room temperature. After the fibronectin solution is removed by aspiration, EC are seeded at 10 to 20 thousand cells per well in M199 growth medium (described above) and allowed to grow for 24 to 48 hours. The confluent EC are washed twice with assay medium (RPMI + 1% FCS) and activated with 100 µl/well cytokine (125 U/ml tumor necrosis factor-α [TNF-α], 2×10^7 U/mg) (Cetus); 10 U/ml interleukin-1β (IL-1β), 5×10^8 U/mg (Genzyme Corp., Boston, MA); or 2.5 µg/ml lipopolysaccharide (LPS) from *Escherichia coli* J5 (Sigma) for 4 hours at 37°C.

Approximately 10^8 PMN, isolated from fresh blood as described above, are radiolabelled with 100 µCi ^{111}Indium oxine (10 µCi/ml) (Amersham Corp., Arlington, MA) in 2.5 ml HBSS for 15 minutes at room temperature. The cells are washed twice with HBSS and resuspended in assay medium at 5×10^5 cells/100 µl. One hundred microliters of the ^{111}Indium-labelled PMN suspension are added to each microtiter well. The plates are incubated for 10 to 30 minutes at 37°C (10-15 minutes for PMN and 30 minutes for monocytes). The wells are filled with assay medium, sealed with adherent plastic plate sealers (Dynatech, Inc., Alexander, VA), inverted onto a pad made from modelling compound, and centrifuged at 90 × g for 5 minutes at room temperature to remove unbound PMN. The plastic seal is removed, the plates are blotted onto tissue, and the residual liquid is removed by careful aspiration. Bound cells are lysed by the addition of 70 µl of buffer (20 mM Tris-HCl, 100 mM NaCl, 0.5% NP-40) on ice for 25 minutes. Samples are counted using Skatron filters and vials. Results using this method are very reproducible. The standard deviations for counts from quadruplicate wells are consistently less than 10% and usually less than 5% of the mean values. This assay protocol does not distinguish between adhesion and subsequent transendothelial migration (see Chapters 4 and 5). Results are expressed as:

$$\% \text{ Specific adhesion} = \left(1 - \frac{cpm_{expt} - cpm_{basal}}{cpm_{total} - cpm_{basal}}\right) \times 100$$

expt = experimental
basal = no activation
total = input

LEUKOCYTE/ENDOTHELIAL CELL SHEAR STRESS ADHESION ASSAY

Efforts to achieve an *in vitro* approximation to the situation existing *in vivo* led to adhesion experiments performed under flow conditions. Initially, shear stress studies in parallel-plate flow chambers were conducted to investigate the characteristics of PMN adhesion and migration,[15] the nature of activated-PMN adhesion to EC,[16] and the relative contributions of CD18-independent and -dependent mechanisms of adhesion.[17] A more recent study has reported that at least the same two mechanisms may be operative when suspended EC bind PMN and monocytes under stirred conditions.[18] This was a conjugate pair formation assay using a two-color fluorescence activated cell sorter (FACS) analysis.

LEUKOCYTE/RECOMBINANT PROTEIN ADHESION ASSAY

Most recently, the availability of the cDNA clones for adhesion molecules and the recombinant proteins has allowed investigators to develop more well-defined variations of the assays described above. Essentially, the binding of leukocytes to purified recombinant adhesion molecules or to transfected cell lines expressing surface adhesion molecules has been studied under static or flow conditions (see Chapters 1 to 5). In fact, the cloning of intercellular adhesion molecule (ICAM-2) and vascular cell adhesion molecule (VCAM-1) was accomplished through the transient expression of unknown genes and the subsequent binding of leukocytes (see Chapter 1). Whether or not expression levels are transient or stable should not detract from its potential to help further define the functional role of adhesion molecules.

TRANSENDOTHELIAL MIGRATION

The reader is referred to the information presented in Chapter 5.

DETECTION METHODS FOR ADHERENT LEUKOCYTES

The quantitation of adherent cells has primarily depended on the prelabelling of leukocytes with the radiolabels ^{51}Cr and ^{111}In. Despite the advantage of high specific activity of radiolabelled cells, the obvious drawbacks of hazardous wastes and label-induced clumping have provided enough incentive for other methods to be discovered. These include the prelabelling method with either the covalent modification of cells with fluorescein isothiocyanate[19] or with the intracellular loading of bis-carboxyethylcarboxyfluorescein (BCECF).[20]

The detection of bound leukocytes can be achieved by methods applied after the adhesion step has been completed. These include the labelling of bound cells with a specific monoclonal antibody conjugated to peroxidase,[21] the measurement of cell-associated myeloperoxidase activity,[22] and the determination of vital stain (Rose Bengal) uptake as a measure of bound leukocytes.[23] Finally, a relatively recent study reports using an electronic cell counter to count the number of suspended EC and leukocytes.[24]

REFERENCES

1. English, D. and Andersen, B.R. (1974) *J. Immunol. Meth.* 5, 249–252.
2. Borish, L., Liu, D.Y., Remold, H., and Rocklin, R.E. (1986) in *Manual of Clinical Laboratory Immunology* (Rose, N.R., Friedman, H., and Fahey, J.L., eds.) *Amer. Soc. Micro.*, Washington, D.C., pp. 282–289.
3. Elliott, M.J., Gamble, J.R., Park, L.S., Vadas, M.A., and Lopez, A.F. (1991) *Blood* 77, 2739–2745.
4. Kimani, G., Tonnesen, M.G., and Henson, P.M. (1988) *J. Immunol.* 140, 3161–3166.
5. Lamas, A.M., Mulroney, C.R., Schleimer, R.P. (1988) *J. Immunol.* 140, 1500–1505.
6. Bevilacqua, M.P., Pober, J.S., Wheeler, M.E., Cotran, R.S., Gimbrone, M.A. (1985) *J. Clin. Invest.* 76:2003–2011.

7. DiCorleto, P.E. and de la Motte, C.A. (1985) *J. Clin. Invest.* 75, 1153–1161.
8. Jaffe, E.A., Nachman, R.L., Becker, C.G., and Minick, C.R. (1973) *J. Clin. Invest.* 52, 2745–2751.
9. Maciag, T., Hoover, G.A., Stemerman, M.B. and Weinstein, R. (1981) *J. Cell Biol.* 91, 420–426.
10. Thornton, S.C., Mueller, S.N., and Levine, E.M. (1983) *Science* 222, 623–625.
11. Maciag, T., Cerundolo, J., Ilsley, S., Kelley, P.R. and Forand, R. (1979) *Proc. Natl. Acad. Sci. USA* 76, 5674–5678.
12. Gamble, J.R. and Vadas, M.A. (1988) *Science* 242, 97–99.
13. Beekhuizen, H., Corsel-vanTilburg, A., and vanFurth, R. (1990) *J. Immunol.* 145, 510–518.
14. Charo, I.F., Yuen, C. and Goldstein, I.M. (1985) *Blood* 65, 473–479.
15. Doroszewski, J. and Kiwala, A. (1988) *J. Cell Sci.* 90, 335–340.
16. Gallik, S., Usami, S., Jan, K.M., and Chien, S. (1989) *Biorheology* 26, 823–834.
17. Lawrence, M.B., Smith, C.W., Eskin, S.G., and McIntire, L.V. (1990) *Blood* 75, 227–237.
18. Kuijpers, T.W., Hakkert, B.C., van Mourik, J.A., and Roos, D. (1990) *J. Immunol.* 145, 2588–2594.
19. Lewinsohn, D.M., Nickoloff, B.J., and Butcher, E.C. (1988) *J. Immunol. Meth.* 110, 93–100.
20. Gimbrone, Jr., M.A., Obin, M.S., Brock, A.F., Luis, E.A., Hass, P.E., Hébert, C.A., Yip, Y.K., Leung, D.W., Lowe, D.G., Kohr, W.J., Darbonne, W.C., Bechtol, K.B. and Baker, J.B. (1989) *Science* 246, 1601–1603.
21. Keizer, G.D., Figdor, C.G., and De Vries, J.E. (1986) *J. Immunol. Meth.* 95, 141–147.
22. Bath, P.M.W., Booth, R.F.G., and Hassall, D.G. (1989) *J. Immunol. Meth.* 118, 59–65.
23. Gamble, J.R. and Vadas, M.A. (1988) *J. Immunol. Meth.* 109, 175–184.
24. McFaul, S.J. and Bowman, P.D. (1990) *J. Immunol. Meth.* 130, 171–176.

GLOSSARY

Alphabetical List of Acronyms for Adhesion Molecules

Acronym	Definition	Family
CR3	complement receptor-3	β_2-integrin (CD11b/CD18)
CR4	complement receptor-4	β_2-integrin (CD11c/CD18)
DREG	down-regulated antigen	L-selectin
ELAM-1	endothelial leukocyte adhesion molecule-1	E-selectin
GMP-140	granule membrane protein-140	P-selectin
LHR	lymph node homing receptor	L-selectin
ICAM-1	intercellular adhesion molecule-1	immunoglobulin (CD54)
ICAM-2	intercellular adhesion molecule-2	immunoglobulin
INCAM-110	inducible cell adhesion molecule-110	immunoglobulin (VCAM-1)
LAM-1	leukocyte adhesion molecule-1	L-selectin
LECCAM	lectin, epidermal growth factor, complement regulatory cell adhesion molecule	selectins
LECAM-1,2,3	a. lectin adhesion molecule-1,2,3 b. leukocyte endothelial cell adhesion molecule-1,2,3	L-, E-, or P-selectin
Leu-CAM	leukocyte-cell adhesion molecule	β_2-integrin
LFA-1	lymphocyte function related antigen-1	β_2-integrin (CD11a/CD18)
LPAM-2	murine lymphocyte Peyer's patch adhesion molecule-2	β_1-integrin (CD49d/CD29)
Mac-1	macrophage-1	β_2-integrin (CD11b/CD18)
Mo-1	monocyte-1	β_2-integrin (CD11b/CD18)
PADGEM	platelet activation dependent granule external membrane	P-selectin

Alphabetical List of Acronyms for Adhesion Molecules *(continued)*

Acronym	Definition	Family
SLe^x	sialyl Lewis X blood antigen	Ligand for E&P selectin
VCAM-1	vascular cell adhesion molecule-1	immunoglobulin (INCAM-110)
VLA-4	very late activation antigen-4	β_1-integrin (CD49d/CD29)

Adhesion Molecule Families	Other Designations	
I. Integrin		
CD11a/CD18	LFA-1	
	$\alpha_L\beta_2$	
CD11b/CD18	Mac-1	CR3
	Mo-1	$\alpha_M\beta_2$
CD11c/CD18	$\alpha_X\beta_2$	
	p150,95	
	CR4	
CD49d/CD29	$\alpha_4\beta_1$	
	VLA-4	
	LPAM-2	
II. Immunoglobulin		
CD54	ICAM-1	
ICAM-2	—	
VCAM-1	INCAM-110	
III. Selectin	LECCAM	
E-selectin	ELAM-1	
	LECAM-2	
L-selectin	LAM-1	Mel-14
	LECAM-1	Leu-8
	gp90^{MEL}	TQ1
	LHR	DREG-56
P-selectin	CD62	
	LECAM-3	
	GMP-140	
	PADGEM	

Index

References to diagrams are followed by *d;* references to figures are followed by *f;* references to tables are followed by *t.*

Activation, 23*t,* 69*f,* 77*f,* 154*f,* 155*t,* 159*t,* 163*f,* 185, 185*f,*186*f*
Activation epitopes of integrins, 164–170, 167*t*
Adhesion
  arenas of activity, 184, 34*f,* 103*f*
  cytokines and, 23*t,* 66–75, 77*f,* 87–89, 94–98, 154–155, 155*t*
  detection methods for, 192
  endogenous activators of, 77*f*
  endogenous inhibitors of, 77*f*
  of eosinophils, 9, 23*t,* 69*t,* 76, 106, 131*t*
  ICAM-1/LFA-1 interactions in, 2–7, 92–93, 129, 154–156
  ICAM-1/CD116 interactions in, 6, 96
  ICAM-2/LFA-1 interactions in, 5–6
  in vitro assay of, 83, 85–86, 94, 97–98, 99, 189–193
  in vivo models of, 33–37, 55–57, 98–101, 130*t,* 131*t,* 136
  inhibition of. *See* Anti-adhesion therapy
  of lymphocytes, 7–8, 10–11, 13, 44–46, 77*f,* 104–106, 151, 160–161
  of monocytes, 23*t,* 32, 34*f,* 68, 69*t,* 76–77, 131, 132, 151
  of neutrophils, 23*t,* 32, 34*f,* 69*t,* 71–76, 103*f*
  receptor-ligand pairs for, 68*t,* 153*f*
  redundancy of pathways, 173–174, 184
  regulation of, 58–60, 152, 154*t*
  selectin/CHO interactions in, 24–28, 27*f,* 32–33, 51–53, 68*t*
  and transendothelial migration, 89–98
  VCAM-1/VLA-4 interactions in, 9–13, 76, 105, 132, 156, 174
Adhesion cascades, 174–175
Adhesion receptors
  acronyms for, 195
  activation of, 68*t,* 159–170
  apparent redundancy of, 173–174, 184
  chimeric, 52–53, 186
  of endothelial cells, 4*f,* 23*t,* 69*t,* 155*t*
  expression of, 23*t,* 69*f,* 155*t,* 185*f,* 186*f*
  families of, 1–2, 3*f,* 4*f,* 20*f,* 196
  genetic factors affecting, 133, 158
  infectious agents affecting, 158
  "inside-out" signalling of, 159–170, 159*t*
  of leukocytes, 3*f,* 23*t,* 69*t,* 155*t*
  location of, 172–173
  "outside-in" signalling through, 159–170, 172
  regulation of expression of, 58–60, 153–158, 154*t*
  soluble, 10–11, 21–22, 49–50, 52–53, 75–75, 186
$\alpha_4\beta_1$. *See* VLA-4
$\alpha_L\beta_2$. *See* LFA-1
$\alpha_M\beta_2$. *See* CD11b/CD18
$\alpha_X\beta_2$. *See* CD11c/CD18

Anti-adhesion therapy, 133–134
  efficacy of, 134–141
  experiments in, 136–137t
  safety of, 141–142
Assay, in vitro, 83, 85–86, 94, 97–98, 99, 189–193
Atherosclerosis. *See* VCAM-1

B cells, migration of, 2, 12, 13, 44–45
Bacterial lipopolysaccharide, 36–37
  in inflammation, 66
β_1 integrins, 8, 195–196. *See* VLA-4
β_2 integrins, 195–96. *See* CD11(a,b,c)/CD18
Burkitt's lymphoma, affecting adhesion, 158

Calmette-Guérin bacillus, 158
CD11/CD18 complex
  deficiency in animals of, 133
  anti-adhesion therapy against, 133–144
  role of, 2, 123, 120–122, 126–129
  and transendothelial migration, 91–92, 94–95
CD11a/CD18. *See* LFA-1
CD11b/CD18
  Mac-1 (Macrophage-1), 195–196
  adhesion mediated by, 6, 96, 126, 173–174
  structure of, 3f
  and transendothelial migration, 91–92
CD11c/CD18
  and transendothelial migration, 95
CD18 integrins. *See* B2 integrins and CD11/CD18 complex
CD2/LFA-3 receptor pair, 156–157
CD43, function of, 163
CD44, function of, 13, 163
CD49d/CD29. *See* VLA-4
CD54. *See* ICAM-1
CD62. *See* P-selectin
C5a, 89, 90
Chimera. *See* adhesion receptors
Cold injury, 141, 142f
Coxsackievirus, ICAM-1 and, 4
CR domain, 21, 48
CR3 (Complement receptor-3). *See* CD11b/CD18
CR4 (Complement receptor-4). *See* CD11c/CD18
Cytokines, 154–158
  activation of endothelial cells by, 1, 66–68, 87–88
  activation of leukocytes by, 70–77, 88–89
  adhesion molecules and, 23t, 67–68, 94–98
  inhibiting endothelial activation, 68–70
  regulation of cell adhesion by, 58–60, 77f, 154–155, 155t
Cytoskeleton, role of, 172–173

D33C MAb, 167t, 170
D3GP MAb, 167t
DREG (Down-regulated antigen). *See* L-selectin
DREG-56. *See* L-selectin

EBV, in adhesion molecule expression, 158
EGF domain, 21, 48
ELAM-1, 157, 195, 196
  activators of, 33
  and adhesion, 7, 32
  carbohydrate ligand of, 25–28
  cell types bound by, 23t, 32–33, 69t
  cytokine effect on, 67–68, 155t, 157
  effect on leukocytes, 69t
  expression of, 22, 23t, 69f, 184
  in inflammation and disease, 22, 36–37
  in lymphocyte-endothelial binding, 7, 32, 105, 174
  and *in vivo* models, 33–36
  ligands of, 25–28, 27f, 30, 32–33
  and neutrophil margination, 101–102
  properties of, 23t
  role of, 131–132
  SLe^x as ligand for, 25–26
  SLe^x expression and, 29–30, 32–33
  structure of, 20–21
  and transendothelial migration, 96–98
  Vim-2 as ligand for, 26–28
Emigration, 125–126, 130t
Endothelial activation
  by cytokines, 66–68, 77f, 87–88
  by histamine, 23, 23t, 33, 35, 102
  by hydrogen peroxide, 23–24, 33, 35
  inhibition of, 68–70, 77f
  by thrombin, 23, 23t, 33, 35, 102
Endothelial cells
  culture of, 190
Endothelial-leukocyte adhesion molecule-1. *See* ELAM-1
Endotoxin. *See* Bacterial lipopolysaccharide
Eosinophils
  adhesion of, 9, 23t, 69t, 76, 106, 132
  in vitro purification of, 189
  migration of, 106, 131t
  SLe^x expression on, 32
Epstein-Barr virus, in adhesion molecule expression, 158

Erythrocytes, size and deformability of, 124
E-selectin. *See* ELAM-1
Extravasation. *See* emigration

Fibronectin, 8, 9, 156, 171
FMLP, adhesion activated by, 90, 91, 94, 159, 162

Glycosylation, 28–32, 152
GMP-140 (Granule membrane protein-140). *See* P-selectin
gp90^{MEL}. *See* L-selectin
granulocytes. *See* Neutrophils

Hemorrhagic shock.
  fluid requirements following, 141f
  visceral organ injury following, 140f
  See Ischemia-reperfusion injury
Histamine, 23, 33, 35, 102, 132
Homing
  genetic background of, 50–51
  lymphocyte, 44–45
  receptor for. *See* L-selectin

ICAM-1 (Intercellular adhesion molecule-1), 195, 196
  cloning of, 2–3
  cytokine effect on, 67
  effect on leukocytes, 69t
  expression of, 69f, 154
  interactions with CD11b/CD18, 6, 96, 126, 173–174
  interactions with infectious agents, 4
  interactions with LFA-1, 2–7, 92–93, 129, 154–156
  in vivo role of, 129
  structure of, 3–4, 4f, 67
  and transendothelial migration, 92–93, 95–96, 105
ICAM-2 (Intercellular adhesion molecule-2), 195, 196
  cloning of, 192
  interactive with LFA-1, 5–6
  role of, 129
  structure of, 4f
Immunoglobulin family, 1
  characteristics of, 2–3
  designations of, 196
  interaction with integrins, 2. *See also* ICAM-1,2 and VCAM-1
INCAM-110. (Inducible cell adhesion molecule-110). *See* (VCAM-1)

Inflammation
  acute, ELAM-1 expression in, 36–37
  adhesion patterns in, 184
  cell movement during, 68
  chronic, expression of ELAM-1 in, 22, 37
  endothelium in, 33–37, 98–101, 122–126
  importance of, 43
  leukocytes in, 152
  lymphocytes in, 165
Integrin receptors. *See* CD11 (abc)/CD18 and VLA-4
  activation of, 163f
  affinity modulation of, 162–163
Integrins, 1
  activation of, 162
  activation epitopes of, 164–170, 167t
  designations of, 196
  interaction with immunoglobulins, 2
  ligands of, 172
  mediating cell-cell interactions, 159
  role of, 171
Intercellular adhesion. *See* Adhesion
Interleukin-1 (IL-1)
  affecting endothelial adhesion, 1, 66–68, 87–88
  affecting neutrophil movement, 87
Interleukin-4 (IL-4), 11
  function of, 12, 70
  affecting endothelial adhesion, 11–12, 77f, 68–70
Interleukin-8 (IL-8)
  affecting neutrophil adhesion, 66, 72–75, 73f, 74f, 77f
Ischemia-reperfusion injury,
  adhesion in, 135, 137–141, 139f
  monoclonal antibodies to treat, 134

L16 MAb, 164–170, 167t
L25 MAb, 170
LAM-1. *See* L-selectin
LECAM-1 (Leukocyte endothelial cell adhesion molecule-1). *See* L-selectin
LECAM-2 (Leukocyte endothelial cell adhesion molecule-2). *See* ELAM-1
LECAM-3 (Leukocyte endothelial cell adhesion molecule-3). *See* P-selectin
LECCAM. *See* Selectins, P- and L-selectin, and ELAM-1
Lectin adhesion molecule-1. *See* L-selectin
Lectin adhesion molecule-2. *See* ELAM-1
Lectin adhesion molecule-3. *See* P-selectin
Lectin domain, 20–21, 47, 48

Lectin, epidermal growth factor, complement regulatory cell adhesion molecule. *See* Selectins
Leu-8. *See* L-selectin
Leukocyte adhesion deficiency (LAD), 43–44
  deficiency in humans of, 118, 124, 125, 126, 133
Leukocyte adhesion molecule-1. *See* L-selectin
Leukocytes
  activation by cytokines, 70–77, 80–89
  adhesion patterns of, 69t, 151, 153f
  endothelial adhesion proteins and, 23t, 69t
  expression of SLe^x on, 29–30
  extravasation of, 103–104, 103f
  interactions with endothelial cells, 33–36, 65
  in marginal pool, 124
  isolation of, 189
  measurement of interactions with endothelial cells, 65–66. *See also* Assay
  migration of, 104–106, 125–126. *See also* Transendothelial migration
  inhibited by monoclonal antibodies, 128–129, 130–131t, 131
  mobility of, 43
  selectins mediating adhesion of, 19–37
Lewis X (Le^x), 25–26
  as antigen, 28–29
  ligand for P-selectin, 28
  structure of, 25d
LFA-1 (Lymphocyte function associated antigen-1), 154–156, 195,196
  activation of, 159–162, 174
  binding to ICAM-1, 2–7, 92–93, 129, 154–156
  epitopes of, 164, 165–166, 166f
LFA-1/ICAM-1 interaction, 154–156
  in initiation of immune response, 173–174
  binding to ICAM-1, 2–7
  and transendothelial migration, 92–93, 95–96, 105
LFA-1/ICAM-2 interaction
  binding to ICAM-2, 5–6
LHR (Lymph node homing receptor). *See* L-selectin
LIBS1 MAb, 167t, 169
LPAM-2. *See* VLA-4
L-selectin, 117, 157, 195, 196
  and adhesion to cytokine-stimulated endothelial cells, 35–36, 71
  carbohydrate ligands of, 24–25
  cDNA cloning of, 46–50
  domains of, 48, 54–55, 184
  expression of, 22
  genomic structure of, 50–51
  and leukocyte rolling, 35, 36, 55–57, 99–101, 123
  leukocyte recruitment affected by, 33, 35, 129–131
  ligands for, 51–54
  as a ligand for ELAM-1, 36, 101–102, 157
  and lymphocyte migration, 44–45, 104–105, 157
  and regulatory aspects of function of, 58–60, 72–73
Leukotriene B4 (LTB4), 89–90
Lymphocytes
  adhesion mechanisms of, 7–8, 10–11, 23t, 45–46, 69t, 104–106, 131, 132, 151–152, 153f, 165
  aggregation of, 9, 160–161, 161f
  emigration of, 44–45, 104–106, 132
  endogenous regulators affecting, 77f, 154–155, 157
  homing ability of, 44–45
  production of, 13

Mac-1 (Macrophage-1). *See* CD11b/CD18
Malaria, ICAM-1 and, 4
Marginated pool, 123–125
Margination, and neutrophil localization, 98–103
Mel-14. *See* L-selectin
Microscopy, intravital, 35–36, 56–57, 99–100, 118–125, 121f
Mo-1 (Monocyte-1). *See* CD11b/CD18
Monoclonal antibodies
  detecting integrin activation epitopes, 164–170
  inhibiting emigration, 128–129
  inhibiting vascular and tissue damage, 134–141
  See anti-adhesion therapy, emigration
Monocytes
  adhesion mechanism of, 23t, 68, 69t, 76–77, 131, 132, 151
  endogenous regulators affecting, 70, 77, 77f
  in vitro purification of, 189
  SLe^x expression on, 28–30, 32
Multiple organ failure syndrome, 138

Neutrophils
  adhesion mechanism of, 23t, 34f, 69t, 71–76, 103–104

endogenous regulators affecting, 77f, 89, 90
in vitro purification of, 189
SLe^x expression on, 28–30, 32
Nitric oxide (NO), inhibiting adhesion, 184
NKI-L16. *See* L16

p150,95. *See* CD11c/CD18
P41 MAb, 167t
PAC-1 MAb, 167t, 169
PADGEM (Platelet activation dependent granule external membrane). *See* P-selectin
Phorbol ester. *See* PMA
PKC, 162–163
Platelet activating factor (PAF), 35, 76, 89, 90
PMA, 6, 23, 92, 97, 127, 132, 159, 173
  causing T cell aggregation, 93, 160, 161t
  induction of phosphorylation by, 162–163
PMI 1 MAb, 167t
PMI 2 MAb, 167t
Polymorphonuclear cells. *See* neutrophils
Protein kinase C. *See* PKC
P-selectin, 157–158, 196
  activators of, 23, 33, 35, 102, 132
  alternative splicing of, 21, 49–50
  cell types bound by, 23t, 69t
  effect on neutrophils, 75–76
  expression of, 22–24
  and leukocyte rolling, 34–35, 102–103, 132
  ligands of, 24–25, 27f, 28, 30, 32–33, 68t, 183–184
  properties of, 23t
  role of, 75–76, 132
  SLe^x expression and, 32–33
  soluble forms of, 21–22, 49–50, 75–76

Rhinovirus, ICAM-1 and, 4
Rolling, 33–34, 98–99
  and adherence, 122–123
  factors affecting, 122
  inhibition of, 99–100
  promotion of, 102–103
  selectin receptors in, 35–36, 55–57, 123

Selectins. *See* ELAM-1, P- and L-Selectin
  in adhesion cascades, 19–37, 157–158, 174
  carbohydrate ligands of, 25–28, 30, 32–33, 51–54, 183–184
  designations and types of, 19–21, 44, 196
  domains, 20–21, 54–55, 184
  expression of, 22–24

Sepsis, anti-CD18 antibody counteracting, 184
Shear stress studies, 191
Signalling
  "inside-out," 159–170, 159t
  "outside-in," 170–172
62 MAb, 167t
SLe^a (Sialyl Lewis A blood antigen), 68t, 183–184
SLe^x (Sialyl Lewis X blood antigen), 196
  biosynthesis of, 30–31, 31d
  distribution of, 28–30
  expression on leukocyte glycoproteins and glycolipids, 29–30, 32–33
  as ligand, 25–26, 28, 32–33, 36, 101–102, 157, 183–184
  structure of, 25d
Soluble receptor. *See* adhesion receptors
Sticking
  CD11/CD18 complex in, 120–122
  discovery of, 118–120
  molecular basis of, 33–34, 34f, 103f, 120, 184f

T cells
  adhesion mechanisms of, 7–8, 10–11, 45–46, 105, 131, 132, 151, 152, 165
  aggregation of, 9, 160–161, 161f
  migration of, 44–45, 104–106
  homing to skin, 44–45, 132, 174
Talin, 172
TASK MAb, 170
Thrombin, 23, 33, 35, 102, 132
TQ1 MAb. *See* L-selectin
Transendothelial migration, 84f, 85f, 184–185
  adhesion and, 89–98
  as a stage of extravasation, 103–104
  induced by endothelial activation, 87–89
  induced by leukocyte activation, 74, 89–91
  in vitro, 83–86
  margination and, 98–103
Transforming growth factor (TGF-), 69–70, 77f
Tumor necrosis factor-α (TNF-a)
  affecting neutrophil adhesion, 66, 71, 77f, 87
  affecting endothelial adhesion, 2, 8, 11, 22–28, 66, 67, 74, 77f
24 MAb, 166, 167t, 168

VCAM-1 (Vascular cell adhesion molecule-1), 156, 170, 196
  alternative splicing, 9–10

cloning of, 8, 67, 192
cytokine effect on, 8, 11–12, 67, 70, 155*t*
and eosinophil/basophil migration, 9, 76, 106, 132
expression in disease, 10, 12, 132
interactions with VLA-4, 7–13, 132
and mononuclear migration, 76, 105, 132
soluble form of, 10–11
structure of, 4*f*, 8
tissue distribution of, 12
Vim-2, as ligand, 26–28
Viruses, binding to ICAM-1, 4
VLA proteins. *See* β_1 integrins
VLA-4 (Very late antigen-4), 196
 activation of, 159
 aggregation, 8–9, 170
 expression of, 156
 interactions with VCAM-1, 7–13
 interactions with fibronectin, 8, 9, 156, 171
 in lymphopoiesis, 13
 role in adhesion, *See* VLA-4/VCAM-1 interactions, 132–133
 as signal transducer, 171
 structure of, 3*f*
VLA-4/VCAM-1 interactions, 156
 in B cell-stromal cell binding, 10, 11, 12, 13
 in eosinophil/basophil-endothelial binding, 9, 69*t*, 76, 106, 132
 in monocyte-endothelial binding, 10, 69*t*, 76, 132
 in lymphocyte-endothelial binding, 10–11, 13, 69*t*, 76, 105, 132, 174

DATE DUE FOR RETURN

CANCELLED

- 3 MAY 2002

1 - APR 2005